Agent Coordination Mechanisms for Solving a Partitioning Task

Andreas Goebels

Dissertation
in Computer Science

submitted to the

Faculty of Electrical Engineering,
Computer Science and Mathematics
University of Paderborn

in partial fulfillment of the requirements for the degree of

doctor rerum naturalium
(Dr. rer. nat.)

Paderborn, October 2006

Bibliografische Information der Deutschen Nationalbibliothek

Die Deutsche Nationalbibliothek verzeichnet diese Publikation in der
Deutschen Nationalbibliografie; detaillierte bibliografische Daten sind
im Internet über http://dnb.d-nb.de abrufbar.

ISBN 978-3-8325-1483-9

Logos Verlag Berlin
Comeniushof, Gubener Str. 47,
10243 Berlin
Tel.: +49 030 42 85 10 90
Fax: +49 030 42 85 10 92
INTERNET: http://www.logos-verlag.de

Abstract

Multi Agent Systems and Swarm Intelligence are two recent and very promising topics in current computer science research. Swarm intelligence deals with large sets of individuals or agents that are regarded as a self-organizing system showing emergent behaviour. Ideas from biology are often and successfully applied to (optimization) problems in the computer science area. Nature provides several examples of complex architectures that are created by very simple insects with highly limited abilities. These insects live in social colonies and coordinate their actions by a concept called stigmergy.

A frequently occurring question when designing multi agent or swarm systems that are (partly) inspired by natural examples is how to coordinate a large group of individuals or instances. This matter is closely connected with the question about the essential characteristics and parameters for both the whole system and each single agent.

The thesis on hand deals with these fundamental questions of multi agent and swarm intelligence systems. It presents several approaches for the different problems that might arise during system design. As background for all these approaches, a complex optimization problem has been chosen. The *Online Partitioning Problem* (*OPP*) addresses the uniform distribution of a group of agents onto targets under several restrictions, i.e. distance minimization and feature restriction. It is intriguing because it is easy to state but often very difficult to solve. Using such a reference problem allows us to compare the single approaches with each other. Though the presented new approaches deal only with this *OPP*, most of them can easily be adapted to other problems or are general concepts.

Acknowledgements

At the final stage of preparing the documentation of my work, this section gives me the opportunity to express my gratitude to all the people who supported and assisted me.

First of all, I want to thank my advisor, Professor Hans Kleine Büning, who always supported my work and helped me with numerous fruitful discussions. He let me work on my own pace, steering the course of my work only slightly whenever it was needed. Further, I am grateful to Professor Friedhelm Meyer auf der Heide who agreed to review this thesis.

I am also truly indebted to the former and current members of the group *Knowledge Based Systems* especially for their scientific support. Above all I would like to thank Natalia Akchurina, Isabela Anciutti, Heinrich Balzer, Uwe Bubeck, Dr. Matthew Henderson, Dr. Elina Hotman, Oliver Kramer, Dr. Theodor Lettmann, Christina Meyer, and Alexander Weimer. I would also like to especially thank Steffen Priesterjahn from the same group who was a co-author of some papers on which parts of this thesis are based on.

Special thanks I owe to many people mainly for being so friendly and helpful in everyday life. The support of Simone Auinger, Gerd Brakhane, Patrizia Höfer, and the whole International Graduate School Team with Dr. Eckhard Steffen, Astrid Canisius, Andrea Effertz, and Martin Decking provided me a perfect backing for this thesis.

The comments of many friends helped me to improve the readability of this manuscript and to make my time as a graduate student enjoyable. Yvonne Bleischwitz, Bartek Gloger, Gül Gülseven, Sabine Hollmann, Diana Kleine, Britta Klütsch, Monika Oswald, and Philipp Roebrock provided a lot of insightful comments and gave a lot of general suggestions for improvement. My supervision of master and bachelor students taught me a lot. I am grateful for the many discussions we had.

Last but not least, I deeply thank my parents, Margret and Klaus, and my siblings, Anne and Christian, for their support and understanding.

Andreas Goebels
Paderborn, October 2006

Contents

List of Figures

List of Tables

List of Algorithms

1

Introduction

Multi Agent Systems and Swarm Intelligence are two recent and very promising topics in current computer science research. Swarm intelligence deals with large sets of individuals, particles or agents that are regarded as a self-organizing system showing emergent behaviour. Ideas from biology are often and successfully applied to (optimization) problems in the computer science area. Nature provides several examples of complex architectures that are created by very simple insects with highly limited abilities. These insects live in social colonies and coordinate their actions by a concept called stigmergy.

A frequently occurring question when designing multi agent or swarm systems that are (partly) inspired by natural examples is how to coordinate a large group of individuals or instances. This matter is closely connected with the question about the essential characteristics and parameters for both the whole system and each single agent. Since the size and likewise the complexity of such systems is constantly rising, it becomes more and more important. At the same time, the capabilities of agents are highly different. This is particularly noticeable if software and hardware systems are considered.

One keyword in this context is the term *emergence*. It describes a system which is as a whole more than the sum of the parts it is composed of. One domain of agent coordination tasks is the analysis of such behaviours. The emergent effects are not necessarily desired, but if they are, they can offer great improvements to the system. For a given problem, several questions regarding the decentralised coordination of agent-interactions arise.

- What are the minimum abilities the agents should have ?
  What type of and how much communication / interaction is needed ?
- How can the individual agents learn an appropriate behaviour ?
- How can the agents organise themselves in useful organisations ?

This thesis deals with these fundamental questions of multi agent system design and presents several solutions for the different problems that might arise. As background for all these approaches, a complex optimization problem has been chosen. The *Online Partitioning Problem* (*OPP*) addresses the uniform distribution of a group of agents onto targets under several restrictions, i.e. distance minimization and feature restriction. It is intriguing because it is easy to state but often very difficult to solve. Using such a reference problem allows us to compare the single approaches with each other.

With regard to different agent abilities, solutions close to the optimum could be found for a wide variety of instances of the *OPP*. Though the presented and successful approaches deal only with this particular problem, most of them can easily be adapted to other problems or are general concepts.

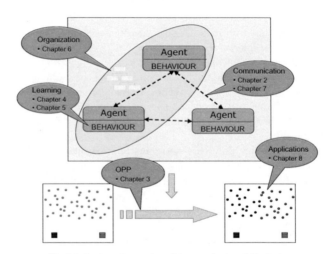

Fig. 1.1. A schematic overview of the organization of this thesis

1.1 Organization of this Thesis

This thesis is organised as follows. Chapter 2 familiarises the reader with the areas that are related to the above mentioned problem fields. Concurrently, a literature overview about research in similar areas is presented. General notations will be formally defined. The second part of this chapter introduces a special graph structure, the k-neighbourhood graph. This graph is highly applicable for communication in large systems as we are dealing with. Some characteristics will be proven and selected properties will be examined in empirical studies.

Chapter 3 defines the optimization problem which all the subsequent approaches try to solve. The *OPP* will be formally defined and compared with similar problems from other research fields. Some complete algorithms will be presented.

The next sections offer heuristics with emphasis on different objectives of the *OPP*.

In chapter 4, basic solution strategies that partly refrain from using any kind of communication or inter-agent knowledge exchange, are introduced. They are used as reference values in the more complex approaches. In the same chapter, we additionally consider a different problem setting. The basic approaches will be analysed with regard to moving agents.

The approach in chapter 5 combines cellular automata research and evolutionary algorithm learning to a heuristic that can deal with instances for the OPP. Initially, the demonstration of a mapping function that can transform arbitrary multi agent systems into assignments for cellular automata is presented. Then, evolutionary algorithms are used to find or learn useful transition rules and, in the last part, the learning algorithm will be improved by introducing some new operators. Concepts from the field of memetics can be noticed throughout this section.

Following this, we convey the multi agent systems into the economic area. The single agents in the system can form organizations which are either able to calculate good partial solutions of an OPP instance or have some advantages through local characteristics. An idea how a recursively structured organization based on only local interactions can be created is presented and examined in chapter 6.

In most of the approaches applied before we use different forms of local communication between agents to enhance the overall result. In chapter 7, the most appropriate communication structure will be discovered by a machine learning process. Based on the k-neighbourhood graphs, introduced in chapter 2, a function is learned that assigns an ideal number of communication partners to any position in space.

In order to illustrate practical aspects of this work, possible applications for the OPP optimization problem are presented in chapter 8. We introduce an algorithm that enhances the data transfer rate in WLan networks by local interactions among the clients. Additionally, we consider the dynamic task allocation problem which is also inspired by natural insect behaviours. The OPP can in addition help to improve the results significantly.

We complete this work with a conclusion that provides a summary of the main results of the thesis. Some future research directions will be mentioned.

The red line of this thesis is visualised in figure 1.1. All areas and the corresponding chapter where these areas are considered are mentioned there.

1.2 Taxonomy of the approaches

In this thesis, several heuristics will be presented. One striking difference between them is the type of communication the single heuristics exploit. Figure 1.2 classifies all approaches according to this characteristic. On the first level, we distinguish between approaches that use some kind of communication and approaches that do not use communication. The communicative ones can be subdivided into restricted and unrestricted strategies. Strategies that have a fixed number k of communication connections to the nearest agents or to its 8 nearest neighbours in the cellular automata

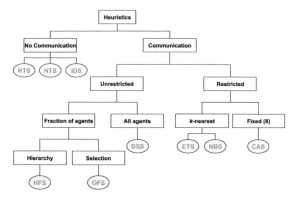

Fig. 1.2. Classification of the introduced heuristics regarding the inter-agent communication type. The abbreviations in the circles represent approaches introduced in this thesis.

approach are restricted. Strategies which use global communication among all agents or among a selection of agents are unrestricted.

1.3 Empirical Analysis

The main focus in this thesis lies on systems with a preliminary unknown but large number of autonomous individuals. Thus, we evaluate most of the heuristics in an empirical way. We want to abstract from the overall number of individuals. In those approaches with only local inter-individual interactions we show that the number of individuals has no influence on the solution quality. For those approaches which contain some global properties, we will discover relations between the number of individuals and the quality of the solution, if they exist.

Additionally, in several approaches we have to find a compromise between the number of learning phases that might further enhance the solution quality and the computing power and time we need for such simulations. The same holds for the iteration of single simulation runs with identical parameters to determine the statistical deviations. In these empirical tests we compare the behaviour and the outcome of the single algorithms for random inputs. In most settings, the agents are randomly and uniformly distributed in parts of the simulation space, but in some settings we examine the behaviour of our algorithms for non-uniformly distributed agents.

2

Definitions and Communication Graphs

The beginning of wisdom
is a definition of terms

Socrates

In this chapter those terms central to this thesis are explained. A summary based on classic and current research is provided. We do not consider border areas of these terms but we will clarify the main ideas.

Following the introduction of terms, a graph structure is presented in section 2.3. The characteristics of these neighbourhood graphs will be examined. We will prove an upper bound for incoming communication connections in two- and three-dimensional space which is each independent from the number of agents. In empirical studies we show that this communication structure has advantages compared to radial limited communication.

The main theoretical results for these neighbourhood graphs have been published in [Goe06c].

2.1 Terms and Notation

In this section, several terms which are used in the remaining part of the thesis will be defined and described. These definitions are introduced independently of the specific optimisation problem considered in the thesis. This is to better place this research into a broader context.

2.1.1 Agent

In commonplace English, 'agent' is described as a person or thing which is able or authorised to act for third parties. The term is derived from Latin, present participle of *agere* (to drive, to lead, to act, to do). The Greek origin is the word *agein* (to drive, to lead) [Dic06]. In the computer science area, this definition is more specified, but a huge number of different definitions and combinations of definitions for the term 'agent' exist. The term is associated with both hardware and software entities. Applying a commonly used approach, we define the term 'agent', as used below, by defining criteria an agent should have. This is mainly based on the books from Caglayan and Harrison [CH97] and Wooldridge [Woo02]. From the first one, the general definition describes a software agent as

a software entity, which fulfils tasks delegated by the user.

The following attributes can be derived:

Delegation: The agent can act according to users behalf. According to [Mae94], the assistance can be realised by training or teaching guidance from the user, by providing inter-user collaboration or by monitoring events and procedures[1]

Ability to communicate: The agent is able to communicate with the user. This communication can be realised via a user interface or any other kind of access.

Interaction with the environment: The agent has actors to change or modify the environment. Wooldridge combines this and the previous point in [Woo02] to *Social Ability*, which is the ability to communicate with users and other agents. In this definition he considers the other agents as part of the environment. For that reason, we distinguish between human users and agents.

Ability to monitor the environment: The agent has sensors or other interfaces to gain information about the environment. Other agents are part of the environment.

Intelligence / Pro-Activeness: The agent has skills to interpret the recognised environment and can act in an appropriate way to fulfil its task. It can exhibit goal-oriented behaviour by taking the initiative.

Autonomy: Finally, the agent can act on its own, based on sensor input and without any need of interaction. [KJ97] and [Wei99] mention the desired abilities of agents to find a way to fulfil the task on their own.

[Woo02] adds several temporal aspects to the definition. He mentions *Reactivity* and *Temporal Continuity*. The first property is an extension of *Interaction with the environment*. In Wooldridge's definition the time for an agents' reaction to changes in the environment is limited. The second aspect assures that an agent is 'alive' for a long period in time.

Agents can be divided into several groups reflecting different abilities. The first and thus most simple agent type is the *Reflex* agent. Agents from this group can be subdivided into table-lookup, simple reflex and reflex agents. All of them solely react to

---

[1] We researched some user imitation ideas in the game / simulation environment *Quake* in [PKWG05, PGW05, PKWG06].

sensor inputs, internally represented in different ways. The next group are the *Goal-Based* agents, these act similar to a greedy hill-climbing algorithm, i.e. they know the overall goal they have to reach. They calculate the outcome of each of the - in the current situation - possible actions and choose the one that brings them closest to the goal. The last group are the *Utility-Based* agents. They can create a plan of action and set up sub-goals. In recent research, agents are not necessarily restricted to a single group. They are successful hybrid solutions composed of different layers. In the same agent, reactive and planning elements can appear at the same time. As a matter of course, through this combination of different layers, new coordination problems can arise. For instance, when the agent is faced with the decision whether an action should be executed by a layer itself or be delegated to the superior layer.

One approach building a layered model has been introduced by Mueller [Mue96]. It is the well known INTERRAP architecture that combines reactive and deliberative components. The lower layer realises local and reactive behaviour while the upper one allows logic deductions in the 'cooperative planning layer'. In the same layer, coordination through communication of information, goals, and plans will be arranged. Ferber [Fer99] provides a different classification. He distinguishes deliberative / reflex behaviours and cognitive / reactive agents. Hence, he has four classes into which all agents can be divided. These are presented in table 2.1.

	Deliberative Behaviour	Reflexive Behaviour
Cognitive Agents	Rational or intentional agents with internal goals. Ability to reason and to perform site-directed actions	Auxiliary agents which react according to requests of other agents
Reactive Agents	Agents which try to satisfy internal goals	Agents which respond to and execute requests, which are the environmental stimuli (tropistic). They neither have internal states nor goals.

Table 2.1. Agent classification according to Ferber [Fer99]

2.1.2 Distributed Artificial Intelligence

Algorithms in the area of Distributed Artificial Intelligence (DAI) have one basic common principle. The whole initial problem $\mathcal{P}$ is split into several (independent) subproblems $\{\mathcal{P}_1, \mathcal{P}_2, ...\}$ which are solved separately. This can happen sequentially. The single solutions will then be combined to a solution for the original problem $\mathcal{P}$. Both, multi agent systems and swarms can be seen as problem solvers that use distributed artificial intelligence.

Multi Agent System

In nearly all applications using agents or agent technology, more than one agent is engaged in the process of problem solving. In such a system with numerous agents, supplementary problems can occur. Each agent has not only to deal with the environment and the problem located in the environment, but in addition it has to cope with the actions of the other agents [Eym00].

In general, agents in multi agent systems act only according to their own well-being.

> *[...] you do not assume that the individual agents have a group sense of utility. Each of the agents in the system can be working at different goals, even conflicting goals. [...] The agents' preferences arise from distinct designers. [RZ94]*

One special research field in the multi agent domain is the so-called *Distributed Problem Solving (DPS)*. There, the whole system, i.e. every agent, is constructed to solely work together with other agents to solve a task. The architecture of each agent is known and there is no egoistic action of the agents which is contrary to the welfare of the whole group [RZ94].

Ferber [Fer99] defines a multi agent system by itemising the elements listed in table 2.2. It can be described by the quintuple (E, O, A, R, Op). One special occurrence

Element	Description
An environment, E	This is a space which generally has a volume
A set of objects, O	These objects are situated, i.e. at any point in time each object can be associated with an explicit position in E.
An assembly of agents, A	This set is a subset of the objects ($A \subseteq O$), representing the active entities of the system.
An assembly of relations, R	Herewith, objects and agents can be linked to each other. Examples are acquaintanceships or communication relations between agents.
An assembly of operations, Op	These operations offer the agents the ability to manipulate objects from O. Manipulation is a generic term subsuming operations like perceiving, creating, consuming, deleting and transforming. So, an operator is a function $op : O \rightarrow O$ that can only be executed by objects that are in A.
Laws of the universe	Extra operators that describe the reaction of the environment to the appliance of these operations.

Table 2.2. Multi agent systems attributes according to Ferber [Fer99]

of such systems is the quintuple (E, A, A, R, Op) with an empty assembly E. In this special case, R defines a network between agents. The network represents the communication structure in a multi agent system. Such systems are very common in distributed artificial intelligence [Yu00].

multi agent systems do not only have the advantages of distributed and concurrent problem solving systems, but some supplementary types of interaction have to be considered additionally. Coordination, cooperation, and negotiation arise as new problem fields. The activities of all agents have to be *coordinated*. In order to achieve that, good interactions are exploited and harmful ones are avoided. *Cooperation* characterises the attitude of the agents to work together to reach a common goal, and *Negotiation* corresponds to the problem of finding an agreement between the agents. The last point is strongly related to game theory research [BPJ02, GvBP00].

> *It is the flexibility and high-level nature of these interactions which distinguish multi agent systems from other forms of software and which provides the underlying power of the paradigm. [Yu00]*

Agent's Intention

In the preceding classification of agent types deducted from Ferber [Fer99] we mentioned that deliberative or telenomic agents have goals of their own which they try to achieve. These goals do not necessarily have to be compatible with each other. If they are, the intention of the agents is of lower importance. In the case that they are not compatible, systems with only altruistic agents can show a totally different behaviour than systems with several or only egoistic players do. Ferber classifies the interactions into 8 groups, depending on the type of goals, the type of resources, and the type of skills.

Agent's Abilities

Stone and Veloso concentrate in [SV00] on the composition of the group of agents. They consider homogeneous systems where all agents are of an identical type. In such a system there is not necessarily a need for communication. With the recursive modelling method (RMM) the authors present a system where an agent can calculate the action of all other agents because it has the same internal structure. Admittedly, this calculation can fail as several deadlock or infinite loop situations exist.

Swarm Intelligence

Emergent behaviour patterns in a set of agents are a widely studied topic in recent computer science research. One current emphasis is on the study of many - very simple - units, instead of few, powerful ones [GP03]. These large sets of simple agents can show very complex group behaviour [BDT99, KE01, Rey87]; examples can be found in lots of variations when considering biological systems [SCW02].

There are several motivations for this relatively new area of research which origins can be traced back to the concept of cellular automata. On the one hand, there are economic reasons. Simple agents, especially when realised in hardware, can be produced with lower costs than big, powerful ones. On the other hand, some philosophical questions occur, for example whether complex behaviour can emerge from interactions of simple units.

To get an impression what 'large / huge number of agents' means, Beni [Ben04] puts it into the order of 10^2 to $10^{<<23}$ agents[2]. He also presented the first definition of this term. Swarm intelligence systems are

> *systems of non-intelligent robots exhibiting collectively intelligent behaviour evident in the ability to unpredictably produce specific [...] ordered patterns of matter in the external environment. Unpredictability is meant as globally 'intractable' or 'externally not-representable'* [BW93]

Typical characteristics of swarms in nature, biology, and computer science are their decentralised control, the lack of synchronicity, and the composition of the group as a set of very simple and (quasi-)identical members (according to [Ben04]). Hence, these characteristics have some significant advantages over other systems. According to [BDT99], the main advantages of swarm intelligence are:

- Scalability: The control architecture of each individual is the same; the overall number of agents does not matter. Due to the strictly local consideration of other members, the overall number of individuals in a swarm is of lower importance. The only limitation is that there are not too few (see the interval proposed by Beni in the last paragraph). Social insects can operate under a wide range of group sizes.
- Flexibility: The individuals / agents may be inserted into or deleted from the environment. There is no requirement for any change in the task operation. One subarea is the (partly or total) failure of single individuals. The whole system can solve the problem provided the amount of lost individuals is below a problem-dependent threshold. Social insects can offer modularised solutions to tasks of different nature by utilising different coordination mechanisms. In nature, the same swarm can fulfil several tasks, for instance 'foraging', 'prey retrieval', and 'chain formation'. [BDT99]
- Robustness / Reliability: The robustness comes from both unit redundancy and minimalist unit design. Social insects can continue to operate despite large disturbances. Some keywords in this area are 'redundancy', 'decentralised coordination', 'simplicity of individuals', and 'distributed sensing' [SS04].

In the following, we will present several areas in the field of swarm intelligence.

Intelligent Swarm

To explain or define what an intelligent swarm is, the term 'intelligence' has to be defined first. There exists a huge amount of literature about this term. Hence, one can say that to deal with the term 'intelligence' is very difficult. Beni [Ben04], who coined the term 'Intelligent Swarm', aims at giving a more detailed insight into it by stating that it is nearly impossible to call something 'intelligent':

> *As is well known, there is no satisfactory definition of intelligence. The concept is elusive. There are many qualities of intelligence, but, for any of them, one can think of some non-intelligent system that has it. [Ben04]*

---

[2] In this thesis, we are dealing mostly with the lower end of this interval. We examine the behaviour of groups of several hundred up to many thousand of agents.

He created many terms for intelligent swarms. In the mentioned paper he designed several definitions and improved them successively. His best choice is that an intelligent swarm is

> *a group of non-intelligent robots ('machines') capable of universal material computation.*

Swarm Robotics

Swarm robotics considers the use of relatively simple robots, equipped with localised sensing abilities, scalable communication mechanisms, and the exploration of decentralised control strategies. Sahin et al. [SS04] describe this field in the preface of the *Swarm Robotics* workshop proceedings with the following words

> *Swarm robotics can be defined as the study of how smart a swarm of relatively simple physically embodied agents can be constructed to collectively accomplish tasks that are beyond the capabilities of a single one.*

Due to the recent technological advances both in hardware as well as in software areas, the study of swarm robotics is becoming more and more feasible. Several projects have been funded in Europe through FET[3] and in the USA through SDR[4] [SS04].

In detail the term 'swarm robotics' is as difficult to formalise as most of the terms in this area are. It can be described to be a multirobot system which consist of large numbers of relatively simple physical robots. The aim of this approach is to study the design of robots (both their physical body and their controlling behaviours) such that a desired collective behaviour emerges from the inter-robot interactions and the interactions of the robots with the environment. This research is inspired but not limited to the emergent behaviour observed in social insects, called swarm intelligence. Unlike distributed robotic systems in general, swarm robotics emphasises large number of robots, and promotes scalability, for instance by using only local communication. This is usually achieved by wireless transmission systems, using radio frequency or infra-red communication. Potential applications for swarm robotics include tasks which demand for extreme miniaturisation such as nano robotics, microbotics, and distributed sensing tasks in micro machinery and the human body. Additionally, swarm robotics is also suited to tasks that demand for extremely cheap designs. For instance, mining or agricultural foraging. Both miniaturisation and cost reduction are hard constraints which emphasise simplicity of the individual team member. [Wik06]

Swarm Optimisation

The *optimisation* aspect has been considered from the beginning of applying swarms in computer science. Best examples are Dorigo's *Ant Colony Optimisation* [CDM92] and Kennedy's and Eberhart's *Particle Swarm Optimisation (PSO)* algorithm [KE01].

---

[3] Future and Emerging Technologies program
[4] Software for Distributed Robotics program

Particle Swarm Optimisation is a population-based stochastic continuous optimisation technique. The approach was inspired by social behaviours of swarms of birds, schools of fish, and, in general, by social cognition theories. There is a high similarity to other nature-inspired approaches, for example evolutionary algorithms (see section 2.1.5). For numerical optimisation, the following holds: In a system with randomly initialised solutions - these are the so-called particles - a global optimum is sought. In each iteration, each particle, located in the solution space, adjusts its position by considering the global best solution and the solution quality of its neighbours.

Each particle has the following properties:

p_{id}: best own position up to now
p_{gd}: best position of all particles up to now
i: the individuality of the particle
s: the social factor of the particle
v: current velocity of the particle. This is a vector which updates the position of the particle in the solution space

The first product of the following summation determines a particle's desire to tend to its own best position found until now. The second product determines its desire to tend to the global best value.

$$\Delta v = p_{id} \cdot i + p_{gd} \cdot s$$

The $updateVelocity()$ function, used in Algorithm 1, can be implemented for parti-

---

Algorithm 1 Particle Swarm Optimisation Algorithm

1: **procedure** PARTICLESWARMOPTIMISATION
2:     $population \leftarrow initialise()$
3:     **repeat**
4:         $population \leftarrow evaluate()$
5:         **for all** (particles p from population) **do**
6:             **if** $(p.fitness > p_{id})$ **then**
7:                 $p_{id} \leftarrow p.fitness$
8:             **end if**
9:         **end for**
10:        $p_{gd} \leftarrow getBestFitness(population)$
11:        **for all** (particles p from population) **do**
12:            $p.updateVelocity()$
13:            $p.updatePosition()$
14:        **end for**
15:    **until** (end condition == TRUE)
16: **end procedure**

---

cle p in the following way:

$$v = v + \alpha \cdot rand() \cdot (p_{id} - p_{fitness}) + \beta \cdot rand() \cdot (p_{gd} - p_{fitness})$$

$rand()$ is a function that calculates uniformly distributed random numbers from the interval $[0; 1]$. The variables α and β are learning factors and are usually set to

$\alpha = \beta = 2$. In line 10, the best value from all particles in the population is chosen. There exist other successful approaches which use solely restricted neighbourhoods to obtain a local best solution.

Until recently PSO had only been applied to single objective problems. However, in a large number of design applications there are a number of competing quantitative measures that define the quality of a solution. In [Fie04], Fieldsend provides a good overview about current approaches and results in this field.

Swarm Engineering

This term has been defined in detail by Alan Winfield et al. in a recent workshop [WHN05]. It considers some problems that might arise during the realisation of swarms. Additionally to the characteristics 'robustness' and 'adaptation', the 'dependability' of the whole system is considered. This direction of research is directed towards analysing how engineered systems based on the swarm intelligence paradigm will be designed, analysed, and tested for dependability in the future.

Kazadi [Kaz00] defines swarm engineering as a formal process by that one creates a swarm of agents which complete a predefined task. He distinguishes between two steps. In the first one, conditions will be generated that guide the generation of a swarm of agents which is capable to fulfil the desired actions. An important point in this step is not to design expressions of the problem with focus on the macroscopic goal, but on the microscopic one. The second step covers the realisation of these conditions. To satisfy them, a set of behaviours will be designed. Kazadi expresses it this way:

> *The goal of swarm engineering is to produce a general condition or set of conditions which may be used to generate many different swarm designs any of which can complete the global goal.*

Swarm Control

This aspect concentrates on the problems that arise when controlling a large set of agents or particles in a swarm. The coordination of the whole group without a central instance is a non-trivial problem. This field includes the problem of coping with asynchronous stability in distributed systems and the formation maintenance issue. An example from the military field dealing with unmanned autonomous vehicles is presented in [Bis03].

Swarms in Nature

The swarm intelligence inspiration is primarily based on models of natural systems, especially in animal societies [Ü93]. Thus, some of the social systems found in nature and researched by the field of swarm intelligence are briefly described in the following.

One of the most organised societies found among animals, ants, are able to select the most rewarding resource, find the shortest path between points, create collective exploratory patterns, deploy together regular nest structures, generate spatial specialisation without requiring communication etc. [Ü93]. By detecting smells and CO_2 they are able to orientate on targets and dispense chemicals to bias other ants to follow their instructions. All colonies have an individual smell that differentiates them from other colonies.

The task organisation in Wasp colonies seems to be a distributed function that does not require a central organiser. Wasps are able to create complex patterns when facing certain types of external constraints. Moreover, wasp colonies present hierarchical (dominant and submissive roles) and tropistic (relation between individuals and the environment) interactions [BDT99]. For instance a honey bee colony manages to react to changes in the food distribution outside the hive and to changes inside the hive, through a decentralised and sophisticated communication and control system [WF05].

> *A honey bee colony can thoroughly monitor a vast region around the hive for rich food sources, nimbly redistribute its foragers within an afternoon, fine-tune its nectar processing to match its nectar collecting, effect cross inhibition between different forager groups to boost its response differential between food sources, precisely regulate its pollen intake in relation to its ratio of internal supply and demand, and limit the expensive process of comb building to times of critical need for additional storage space. [See95]*

The motivation for fish to stay in a group is a bit different. Schooling helps fish to easier find food or to better avoid predators. To keep united, the sensors of fish are basically the vision and the lateral lines; the first seems to provide attractive force, while the latter apparently generates a repulsive force. Regardless its formation abilities, groups of fish can split in smaller clusters in order to improve hunting [Par82].

The behaviour of flocks of birds is similar to the behaviour of schools of fish. They keep themselves together by attractive and repulsive forces, turns are coordinated, and they avoid collisions with each other by very simple rules. Their coordinated behaviour can be explained by assuming that each bird tries to maintain a specified separation from the nearest birds and to match the velocity of nearby birds. [Rey87]

Termites possess the ability to construct huge nests for their entire population. They also use pheromones, thus generating an 'aromatic potential field' [BDT99]. In constructions of termites there may be several orders of magnitude of difference between the size of an individual and the size of a nest built by the colony: for instance, the ratio 'nest size' to 'individual size' may reach 10^4 up to 10^5 in some termite species (for example *Macrotermes bellicosus* [Gra59]). Figure 2.1 shows such a structure. Furthermore, nest structures found in termites often combine a high degree of regularity (for example, nests of the termite *Apicotermes*) with a large diversity of subunits that allow complex regulatory mechanisms to be implemented. For instance, in some parts of the hive are spiral cooling vents[5] and there is an agriculturally used part where the insects cultivate fungi. [BTD⁺98b]

Fig. 2.1. An example of a nest of *Apicotermes* (c) by Masson

2.1.3 Self-Organisation

The term *self-organisation (SO)* (in this context) was first defined in 1954 by Farley and Clark as a system that changes its basic structure as a function of its experience and its environment [YJG62]. Self-organisation is widespread in different (research) areas. In all of these, there are very different concepts and notations. The theory of self-organising systems has first been developed in physical and chemical research areas to explain the occurrence of macroscopic patterns based on processes on the microscopic level. Bonabeau et al. have shown in [BTD⁺97] that this theory can be expanded to ethology, describing partly the behaviour of social insects. They have pointed out that complex group behaviour can emerge from local interactions between simple individuals. The advantage for social animals to act in a self-organising manner is obvious, there is no need for complex behaviours of single individuals. This might be a reason why evolution preferred this kind of cooperation [BTD⁺97]. There are several requirements to a system to be denoted as a self-organising framework. Decker [Dec00] addresses the following indicators for self-organising systems:

---

[5] There have been examinations that individuals can detect air movements one-thousandth the amplitude of those present in a closed room. [BTD⁺98b]

Thermodynamic Open: There has to be an energy flow into the system. This avoids the possibility that the behaviour of the system stagnates due to missing stimulants.

Dynamic Behaviour: The system has to change constantly. Energy from outside of the system has to be converted into entropy that can be observed from outside.

Local Interactions: All parts in the system have to interact with others.

Existence of Multiple Elements: The number of elements in a self-organising system should be very high because the overall behaviour emerges from the interaction of these elements.

Emergence: A complex behaviour has to arise in the system that is more complex than the sum of the single elements. Therefore, there have to exist at least two distinct levels that can be considered: The microscopic and the macroscopic level.

Non-linear Dynamics: A self-organising system needs loops with positive and negative reactions. The interactions should appear between parts on the same level and between parts on different hierarchical levels.

All the points mentioned above are difficult to formalise and it is not guaranteed that a system with all of these properties finally shows a self-organising behaviour, but they present important characteristics of such systems.

In the next sections, several research areas of self-organisation will be presented shortly. More areas and examples can be found for example in [KW97].

Self Organisation in Physics and Chemistry

Equilibrium Statistical Physics (ESP) examines self-organisation phenomena in complex systems between simple physical objects. These objects, often called *particles*, interact only with other nearby objects according to deterministic laws [AM76, Rei65]. ESP tries to find explanations for why and when macroscopic behaviour can arise in such systems. The term *Self Organisation* has first been introduced in these natural sciences.

Self Organisation in Nature

Many phenomena in nature can be explained by a self-organisation process. One important area is the behaviour of social insects. In computer science, this field is called *swarm intelligence*. In section 2.1.2 we concentrate on this part. But there are several other interesting areas. If one considers the development of sand dunes, avalanches or clouds, the self-organisational processes are obvious. When explaining the stigmergy term in section 2.1.4, we return to self-organisation in nature.

Self organisation in Economics

Several areas in economics deal with self-organisation processes. A good overview is presented in [SS98]. In chapter 6 several ideas for an emergent establishment of inter-agent organisation structures are presented. Kirman (in [SS98]) emphasises again the two different levels:

> *There are two essential things to examine: how the organisation of the inter-action between the individuals and the component parts of the system affects aggregate behaviour and how that organisation itself appears*

In this context, Adam Smith's *Invisible Hand* [Smi76] can be seen as a self-organisation process. The wealth of the whole nation is a direct or emergent consequence of the optimisation of the personal situation of each individual in the society. Very often, economic sciences examine one selected individual from society as a representative and deduct from its behaviour to the behaviour of the whole economy. But new approaches model customers or customer-producer relations with self-organising techniques. Therewith, different areas in the market behaviour can be reproduced and explained. Examples are the development of a limited number of production centres, the motivation of regular customers or stable trade connections, the composition of useful communication structures between customers and producers, or the development and fixing of prices. In all examples, the behaviours of a mass of persons or companies act according to local rules and produce therewith a global, macroscopic pattern. More examples can be found in [SS98]. Nowadays, we denote this system behaviour to be self-organising, but one can consider the classic terms *price mechanism*, *equillibrium* and *competition* as some similar areas.

Self Organisation in Sociology

In sociology, several phenomena related to the self-organisation term can be observed. Examples are herd behaviour and groupthinking. A combination of sociology, agent systems, and computer science are the so-called *Agent-Based Social Simulations (ABSS)*. There, emergent effects that occur in social systems will be analysed. In this context, emergence is defined as

> *a system property in which system behaviours at a higher level of abstraction are caused by behaviours at a lower level of abstraction which could not be predicted or estimated at the lower level.* [MD00]

Social systems are in the majority of cases very scalable, robust, and open. Recent applications with connections to this area are P2P software, webservices, and AdHoc-networks. But research in this area will also be used to explain or predict human or, generally speaking, group behaviours.

One famous example is the *Sugarscape* environment from Epstein and Axtell [EA96]. In their approach, a multi agent system is combined with a cellular automaton. Thus,

very complex structures can be observed. The system is able to explain several behaviours that occur in huge societies by the implementation of very few and very
simple rules.

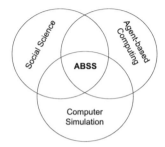

Fig. 2.2. The parts that are involved in agent-based social simulations

2.1.4 Stigmergy

Stigmergy denotes the concept of using the environment as a communication medium. When using stigmergy in the area of multi agent systems, there is no direct
inter-agent communication existent. The agents communicate by sensing modifications of the environment made by other agents. In doing so, information can be transferred among agents. Depending on the environment, with this concept a one-to-one
and one-to-many communication can be realised. Moreover, the communication can
be performed for different periods of time.

The stigmergy information transfer depends highly on the position of the agents and
on the position of the information modification in the environment; for that reason,
the idea of stigmergy can be described by:

The worker [agent] does not direct his work but is guided by it. [HM99]

If the reaction to the environment depends on the type of information obtained from
it, the agent makes use of *Qualitative Stigmergy*. For example, a stimulus of certain
type, type I say, can produce a certain action, A say. With this action the stimulus can
be transformed to a stimulus of *type II*. This again can excite an action B and so on.
In nature, examples for such kinds of stigmergy can be found for instance in nest construction behaviours of colonies of the *polistes dominulus* wasps [BDT99, Goe03].

In contrast, if only the amount or strength of the information or stimulus causes reactions, this is called *Quantitative Stigmergy*. Here, the choice of an appropriate action
depends on the intensity of the stimulus. A stimulus in the interval $[0, a[$ enables action
A, a stimulus in $[a, b[$ (with $b > a$) action B and so on. In nature, this kind of stigmergy can be observed for example in larvae arrangements in wasp colonies. The size
of existing larvae clusters guides the behaviour of larvae carrying wasps. The larger

the existing cluster is, the higher is the probability to drop the load. A similar concept is used as a solution to the dynamic task allocation. This problem is considered in section 8.2 of our application chapter. Another example is the path following or food source exploitation of ants. The behaviour is guided by the pheromone concentration in the environment. The higher the pheromone concentration in one direction is, the higher is the probability for walking in this direction. This pheromone concentration emerges from ants that found a food source. Since the pheromones evaporate by-and-by, the highest pheromone concentration represents the momentary shortest path found by the colony, it is a local optimum. The stochastic behaviour of ants assist the search for new solutions, therefore ants are able to find a global optimum. In most of the existing real-life systems there is no pure qualitative or quantitative stigmergy concept but a combination of both types.

In computer science, the stigmergy concept and swarm algorithms are used for several heuristics handling standard problems. Ant algorithms have been successfully applied to find good solutions for the travelling salesman problem, inspired by food exploitation of ants in nature [DG97]. Similar ideas are used to optimise routing tables in large and dynamic networks. Comparatively new approaches use wasps' or bees' dance language and information transfer ideas to find solutions for the satisfiability problem and to transfer data about the underlying network structure between routers. [WF05]

Stigmergy and Social Insects

Grassè [Gra59] was the first researcher that could explain the complex behaviour of social insects by introducing the stigmergy concept. He could interpret several emergent group behaviours as a result of stigmergy. Two types of behaviour patterns can be distinguished. Insects using the first type are mainly individualists. They have some kind of internal state that can be changed by external stimuli. Hence, these insects can be seen as some kind of state automata with an internal state and transformations executed mainly by external, environmental factors obtained via stigmergy.

The second type does not need an internal state in the insects themselves. A sequence of actions can be achieved using only external stimuli. A shift of an external stimulus depends often on the successful completion of a specific action. Here, the environment acts as some kind of external memory, there is no need for these insects to transfer state information or to communicate directly. This type is used both by individual and social species [Fer99].

Stigmergy and Self-Organisation

Bonabeau et al. [BDT99] and Holland et al. [HM99] describe the attributes and elements of stigmergy in the area of social insects. Herewith, spatio-temporal structures in an originally homogeneous medium can be created (for example pheromone paths or social structures in the group) and the system converges to one of many stable states (multistability). The final state depends on the initial parameters of the system. The last attribute is the origin of the chaotic behaviour of the system; minor changes can result in radically different outcomes. The individual elements are:

Positive feedback: Actions of agents can be strengthened and therefore become more and more attractive over time by introducing a positive feedback. An example for such behaviour is the search for a shortest path to food sources done by several families of ants. The shorter the path is, the higher is the pheromone concentration. And a higher pheromone concentration is proportionally connected with the attractiveness of the path.

Negative feedback: By adding a negative feedback to the system, it becomes more and more stable because herewith wrong decisions can be corrected and the exploration of different solutions can be realised.

Amplification of fluctuations: This element expands the system by adding randomness. If it is not up to the original problem, such characteristics can be realised by random movements, random, incomplete or fuzzy sensor data etc.

Presence of multiple interactions: The stochastic properties can be used by a huge number of different interactions so that the production of big and stable structures is supported.

Self organisation due to stigmergy is rather different from the pure physical self-organisation. Stigmergy SO originates from agents that move according to own, individual calculations. In such systems, structures can be generated on the one hand by the environment itself and on the other hand by the distribution of the agents. Therefore, the overall number of structures is significantly larger. A comprehensive overview about literature using the term stigmergy can be found in [She03].

2.1.5 Evolutionary Algorithms

Evolutionary algorithms (EA) are inspired by the very successful evolution principle found in biology, introduced by Darwin in *The Origin of Species* [Dar59]. EAs can be seen as a search heuristic which is able to deal with extensive search spaces, as nature is one in real life. The basic idea is to select from non-optimal solutions some good ones, recombine and modify them to obtain new solutions and iterate this process until some stop criterion is reached. The Darwinian concept of *survival of the fittest* ensures that during sufficient long runs a good solution can be found by such a search heuristic. The evaluation of the solutions is done by a fitness function. There are three basic operators that guide the search, first the *select* operator which chooses solutions. The selection process can be realised by many different approaches, for example *Best-Selection* that selects the k best intermediate solutions, *Tournament-Selection* or *Roulette-Wheel-Selection*. The latter selects each solution according to its fitness value. There, solutions with a high fitness value have a higher probability to be selected. In the tournament selection, one of two individuals, randomly chosen, will be taken over to the next generation depending on their fitness value in some kind of private competition.

The second operator is responsible for recombining two[6] solutions, often called parents, into one or more new solutions. These new solutions, often called 'children', have properties from both parents. Again, there exist a lot of different approaches for this recombination process, two basic ones are the k-point and the uniform crossover. The last popular operator is *mutation*. With it, solutions will be slightly modified. This operator can either be applied only to children solutions, to all solutions or to any other subset of all current solutions.

The *selection* and *crossover* operators are the exploitation part of the search heuristics, by adding the *mutation* operator the algorithm can explore the whole search space and can escape from local minima. The whole process is generation based, so in each iteration of the loop all instantaneous solutions are managed in a generation [ES03]. There are possibly different stop criteria for the loop starting in line 3 of Algorithm

---

Algorithm 2 Evolutionary Algorithm

```
 1: procedure GENETICSEARCH
 2:     population ← createRandomPopulation()                    ▷ initialisation
 3:     while (not reached end criterion) do
 4:         population ← evaluate(population)
 5:         population ← select(population)          ▷ selection operator for crossover
 6:         population ← crossover(population)                  ▷ crossover operator
 7:         population ← mutate(population)                      ▷ mutation operator
 8:         population ← select(population)            ▷ selection operator for survive
 9:     end while
10: end procedure
```

---

2. The search could end for example when a highest / average fitness value is reached or termination can occur when a maximum number of iterations is reached.

To use evolutionary algorithms for optimisation problems, the solutions have to be encoded in a DNA-like structure. One possible realisation is the bit-string representation, the so-called *genotype*. The *phenotype* is the interpretation of the genotype. The whole string is called a *chromosome* and a single element of a chromosome is called *gene*.

Evolutionary algorithms are universal problem solvers that can cope with complex and non-linear problems. They are mostly not the best, but the second best algorithm for many problems [ES03].

EA Example

Next, we will explain the functioning of an evolutionary algorithm by providing a simple example. Suppose we wish to choose $x_i \in \{0, 1\}$ for all $1 \leq i \leq n$ so as to maximise the value of $\sum_{i=1}^{n} x_i$.

---

[6] or sometimes more, see for example [Tin05]

Obviously, the optimal solution is $x_1 = x_2 = ... = x_n = 1$. For the evaluation of a solution $\mathcal{S} = \{s_1, s_2, ..., s_n\}$, a fitness function can simply sum up all elements in $\mathcal{S}$. The crossover operator can recombine two solutions $\mathcal{S} = \{s_1, s_2, ..., s_n\}$ and $\mathcal{S}' = \{s_1', s_2', ..., s_n'\}$ into two new solutions, for example by applying a one-point crossover [ES03] operation (at point p) we obtain:

$$(\mathcal{C}, \mathcal{C}') \leftarrow crossover(\mathcal{S}, \mathcal{S}', p)$$

with

$$\mathcal{C} = (s_1, s_2, ..., s_p, s_{p+1}', ..., s_n') \text{ and } \mathcal{C}' = (s_1', s_2', ..., s_p', s_{p+1}, ..., s_n)$$

Mutation of a solution $\mathcal{S} = \{s_1, s_2, ..., s_n\}$ can be done by replacing, with small probability, the value s_i with $|s_i - 1|$ for all $1 \leq i \leq n$.

2.2 Problem Notation / Definitions

In this thesis, we denote for $n, m \in \mathbb{N}$ the set of all agents with $\mathcal{A} = \{a_1, ..., a_n\}$ and the set of targets[7] with $\mathcal{T} = \{T_1, ..., T_m\}$.

Definition 1. *When we have a set $\mathcal{A}$ of agents and a set $\mathcal{T}$ of targets we often talk of them as being positioned in space. By this we mean that there exists some injective position function*

$$\rho : \mathcal{A} \cup \mathcal{T} \rightarrow \mathbb{R} \times \mathbb{R}$$

and we speak of the value $\rho(o) = (x, y)$ for an object $o \in \mathcal{A} \cup \mathcal{T}$ as being the position of o.
Analogical, in three-dimensional environments, the function is

$$\rho : \mathcal{A} \cup \mathcal{T} \rightarrow \mathbb{R} \times \mathbb{R} \times \mathbb{R} : \rho(o) = (x, y, z) \quad (o \in \mathcal{A} \cup \mathcal{T})$$

$\rho_x = \rho_1, \rho_y = \rho_2, \rho_z = \rho_3$ *are the coordinate functions of ρ. For instance, $\rho_x(a)$ (and $\rho_1(a)$) return the x coordinate of agent a in the simulation space.*

Definition 2 (Distance Function). *We speak about the distance between two objects $o, o' \in \mathcal{A} \cup \mathcal{T}$ as being the value $\delta(o, o')$ under some mapping*

$$\delta : (\mathcal{A} \cup \mathcal{T}) \times (\mathcal{A} \cup \mathcal{T}) \rightarrow \mathbb{R}_0^+$$

Such a mapping is called a distance function.
For agents located in the d-dimensional space, distance will be calculated by

$$\delta : (\mathcal{A} \cup \mathcal{T}) \times (\mathcal{A} \cup \mathcal{T}) \rightarrow \mathbb{R} : \delta(o, o') = ||o, o'|| = \left(\sum_{i=1}^{d} (\rho_i(o) - \rho_i(o'))^r \right)^{1/r}$$

---

[7] These are elements the agents should assign themselves to.

For $r = 1$, δ determines the Manhattan- or Hamming distance, which is in two-dimensional space equal to:

$$\delta_M(o, o') = ||o, o'||_M = ||\rho_x(o) - \rho_x(o')|| + ||\rho_y(o) - \rho_y(o')||$$

For $r = 2$, δ determines the Euclidean distance, which is in two-dimensional space equal to:

$$\delta(o, o') = ||o, o'|| = \sqrt{(\rho_x(o) - \rho_x(o'))^2 + (\rho_y(o) - \rho_y(o'))^2}$$

The distance can never become negative and these functions fulfil the triangle inequality.

Definition 3. *The* plan *of an agent is the currently chosen target. A plan can change over time. If an agent pursues a plan then it is called* stated. *We use the function τ for the access of an agents' plan. Each agent has a stated plan, a natural number, and the plan of agent a is the image of a under the map*

$$\tau : \mathcal{A} \rightarrow \mathcal{T} : \tau(a) = T_i, T_i \in \mathcal{T}$$

Definition 4. *An agent is* minimum stated *if it has chosen a τ_{min} such that*

$$\tau_{min} : \mathcal{A} \rightarrow \mathcal{T} : \tau_{min}(a) = \min_{T \in \mathcal{T}}\{\delta(a, T)\}$$

2.3 Neighbourhood Graphs

This section specifies properties of a special type of graph, the k-neighbourhood graph. It can be used to establish sparse communication structures in large groups of agents as they are present in swarm applications. We will prove some properties of k-neighbourhood graphs. For example, we show that the maximum number of incoming connections per agent is limited to a value independent of the overall number of agents in the system. We examine the connectivity of random graphs. Then we make an empirical examination of the overall communication energy costs for the whole system with our approach and compare these results with another type of graph which is often used in the area of AdHoc networks. Finally, suitable values for random agent settings derived from experimental trials will be presented. We give an upper and a lower bound for this value.

2.3.1 Introduction

In several multi agent or swarm intelligence systems it is shown that inter-agent communication can improve the performance of the whole system (see for example [Mat98]). Several approaches exist that apply different techniques and concepts to establish such communication structures. These techniques can be divided into two main areas.

The first one can be characterised by methods that offer global communication for each agent in the group. This can be realised by centralised techniques, for example the usage of a black- or whiteboard [Nii86], or by allowing each agent to communicate directly with any other in some kind of complete communication structure [Fer99]. When dealing with large numbers of agents as they are present in swarms or swarm like systems, the centralised approaches have some disadvantages. The blackboard can become a bottleneck and is a single point of failure. The complete communication structure shows some problems in such large settings, too, since the amount of messages grows very fast in such systems.

The second area is more applicable to large groups of agents since it allows only communication in a local area, mostly defined by a maximum communication radius [GGH$^+$01] or by limiting communication to a special direction, for example defined by a cone. Such latter graphs are named after the author and are called Yao graphs [Epp96]. These systems need a fixed, environment- or problem-dependent radius. A radius which is too big will produce similar problems as the global communication concepts entail because of too many communication partners. A radius which is too small can lead to a system with no communication (and hence no information exchange) at all. Other concepts in this area are for example *c-spanners*. These are graphs such that between any two vertices there exists a path of length at most c in geometric distance [SVZ04].

Our approach is related to the local communication concepts. The difference to the ideas we mentioned before is that we allow each agent to adjust its communication radius individually. Compared to Yao graphs, we do not claim that our neighbours are in different sectors. In detail, each agent will increase its radius until a fixed number k of other agents are accessible, wherever they may be located. We model our system by transforming the agents and communication lines into a more abstract geometrical graph structure. In such a graph, each agent is represented by a node and each communication line is an edge. In the following text we will use the terms *node* and *agent* in parallel. The arising graph is denoted as a k-neighbourhood graph (kNG). In chapter 7 an idea for learning appropriate sizes of the k-neighbourhood for each agent individually when handling a multi-objective optimisation problem is presented, but in this chapter the k is identical for all agents.

The remaining part of this section is structured as follows. In the next section, we give a formal definition of the k-neighbourhood graph and introduce some notations. In sections 2.3.3 and 2.3.4, we prove some properties of arbitrary kNGs and will show that the number of communication lines for each agent is independent from the overall number of agents in the system. Sections 2.3.5 and 2.3.6 present experimental results on random settings and show the quality of such graphs according to graph connectivity and energy consumption.

2.3.2 Notation

In this section, we formally define the k-neighbourhood graph and several functions and terms we use in this thesis. It is assume that no two agents are mutually coincident. In other words, we assume that $\delta(a, a') > 0 \ \forall a, a' \in \mathcal{A}$ such that $a \neq a'$.

The k-neighbourhood $\aleph_a$ of an agent a in a group of agents $\mathcal{A}$ is the k-subset S ($S \subseteq \mathcal{A}, |S| = k$) of agents that are nearest k neighbours to a.

Definition 5 (k-neighbourhood). *The k-neighbourhood $\aleph_a$ of an agent $a \in \mathcal{A}$ is defined to contain the k nearest agents.*

$$S = \aleph_a = (a_1, ..., a_k) \text{ such that } \forall a' \in (\mathcal{A} \setminus \aleph_a) : \delta(a', a) \geq \delta(a_i, a) \ (i = 1, ..., k)$$

By definition, the neighborhood $\aleph_a$ does not contain the agent a itself.

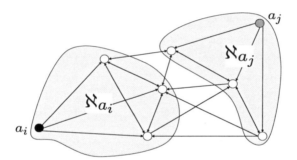

Fig. 2.3. A 3NG with two example 3-neighbourhoods

A *k-neighbourhood graph (kNG)* is a structure that arises when nearby agents connect to each other. In detail, each agent creates connections to its k nearest neighbours.

Definition 6 (k-neighbourhood graph). *We consider the graph $G = (V, E)$ created when connecting each of the n nodes ($n = |\mathcal{A}|$) with their k nearest nodes. These are defined by the k-neighbourhood.*

$$V = \mathcal{A}; E = \{(a, b) | a \in \mathcal{A}, b \in \aleph_a\}$$

We call this graph G a k-neighbourhood graph (kNG). If the edges are directed, the graph is a directed-kNG.

In [Vai89], there is an algorithm presented that can compute such graphs in time $O(kn \log n)$ and in [Cal93] the algorithm needs $O(kn + n \log n)$.

A *k-connected neighbourhood graph (kCNG)* is a kNG which is connected, i.e. there are no independent sub-graphs.

Definition 7 (k-connected neighbourhood graph). *We consider the kNG $G =$ (V, E) created when connecting each of the n agents to its k nearest neighbours with*

$$V = \mathcal{A}, E = \{(a, b) | b \in \aleph_a\} \ \forall a \in \mathcal{A}$$

We call this directed graph G a k-connected neighbourhood graph (kCNG) if for all pairs of agents $(a, b); a, b \in \mathcal{A}$ there exists a subset $P_{a,b} \subset E$ and such that there exists a path from a to b only over edges $\in P_{ab}$.

2.3.3 kNG in One- and Two-Dimensional Space

In this section, we will give some theoretical insight into k-neighbourhood graphs. It will be proved geometrically that the number of incoming connections to an arbitrary node is independent from the overall number of agents and bounded by $2 \cdot k$ (1D) and $6 \cdot k$ (2D). We start with agents located in one-dimensional space, i.e. on a line, and will then proceed to two-dimensional settings.

Theorem 1. *In a one-dimensional space the maximum number of incoming connections to an arbitrary agent $a_M \in \mathcal{A}$ is bounded by $2 \cdot k$ if all agents are connected by a kNG.*

Proof. We consider all agents on the left side of the agent a_M. Assume there are $k' > k$ edges from agents - denoted by $l_1, \ldots, l_{k'}$ - to the agent a_M (see fig. 2.4). Now

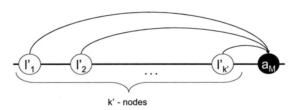

Fig. 2.4. The schematic graphs shows our assumption that k' nodes have chosen agent a_M to be in their neighbourhood.

we sort these agents according to the distance measure $\delta(l_i, a_M)$ such that there are k' agents $l'_1 \ldots l'_{k'} \in A$ with

$$\delta(l'_i, a_M) > \delta(l'_j, a_M) \text{ for } i < j \ \forall i, j \in \{1, \ldots, k'\}, (i \neq j)$$

Therefore we know that

$$\delta(l'_i, l'_{j_1}) < \delta(l'_i, l'_{j_2}) \text{ for } 1 \leq i < j_1 < j_2 \leq k'$$

Now we consider the leftmost agent connected to the agent a_M, this is the agent l'_1. We know that there are at least $(k' - 1) \geq k$ remaining agents between l'_1 and a_M and we know that $\delta(l'_1, l'_m) < \delta(l'_1, a_M)$ for all $m \in \{2, ..., k'\}$. Therefore, a_M is not one of the k-nearest agents to agent l'_1 and there cannot be a connection between l'_1 and a_M. This is a contradiction to our assumption that l'_1 is connected to a_M. In the same manner we can show that no other agent l'_x with $x < (k' - k)$ can have a connection to a_M, therefore k' has to be $\leq k$.

We can argue in the same way for the agents on the right side of a_M and therefore agent a_M can have a maximum of $k + k = 2 \cdot k$ incoming connections from other agents. $\square$

Now, we switch to agents located in Euclidean two-dimensional space. We start with a proof for $k = 1$ and extend this one for arbitrary values of k. In [EPY97] there is another idea for a proof only for $k = 1$ that shows, using characteristics of kissing numbers, the same properties, but we need the approach presented here to show the more general relationship for arbitrary k.

Theorem 2. *In a two-dimensional space the maximum number of incoming connections to an arbitrary agent $a_M \in A$ in a 1-neighbourhood graph is exactly 6.*

We begin with a Lemma proving that it is possible to locate 7 agents in such a way that one agent has 6 incoming connections.

Lemma 1. *For special values of n, agents can be located in such a way that one agent has 6 incoming connections if the agents are connected by a 1-neigbourhood graph.*

Proof. We arrange $(n-1)$ agents on the edges of a regular polygon with the remaining agent a_M in the centre of the polygon (see figure 2.5). We denote the distance between

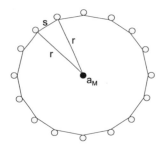

Fig. 2.5. The circle formation of $(n-1)$ agents around an arbitrary agent a_M.

each agent and the agent a_M with r and the distance between two adjacent agents with s. We have the law of cosine

$$s^2 = 2r^2 - 2r^2 \cos\left(\frac{2\pi}{n-1}\right)$$

if and only if

$$r = \pm \underbrace{\left(2 - 2\cos\left(\frac{2\pi}{n-1}\right)\right)^{\frac{1}{2}}}_{=\Delta} \cdot s$$

Here, we can obviously ignore the negative value for the radius r.

For Δ smaller than 1, r is smaller than s. Hence, there will be a connection from an agent in the polygon face with the agent a_M. This inequality $\Delta \leq 1$ is only true for $(n-1) < 6$ or $n < 7$. In the case $n = 7$ the distance from one agent to the agent a_M in the middle of the polygon is the same as the distance to its two adjacent polygon neighbours, therefore the agents can still choose the agent in the middle and the in-degree of a_M can be 6. $\square$

Now we start with the proof for Theorem 2. Therefore, we divide the space around an arbitrary agent a_M in s sectors S_i with equal size (see figure 2.6(a)). We choose for each sector the agent n_i with the smallest distance to agent a_M. An agent a will be in the sector S_i if it is located directly inside a sector. If an agent is located on the border between two sectors, it will belong only to the right one (clockwise seen). Next, we consider an arbitrary sector S_i with the agent n_i that lies nearest to a_M,

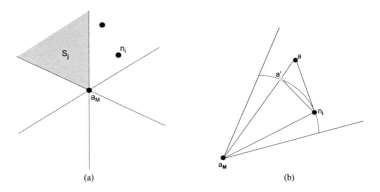

(a)   (b)

Fig. 2.6. The left sketch shows the division of the space into 6 sectors $S_1...S_6$ with agent a_M as origin. The right sketch shows a sketch of one sector that contains n_i and a.

and we show that there can be at most one connection to a_M. Let A_i be the set of agents that are in this sector S_i with $n_i, a_M \notin A_i$. For all agents $a \in A_i$ it holds that $\delta(a, a_M) > \delta(a, n_i)$ if we have a sufficient large number s of sectors. This is proved in the following Lemma.

Lemma 2. *Let S be a sector originating at a_M with an angle α in a two-dimensional space and let n_i and a be two points in S with $||a_M, n_i|| \leq ||a_M, a||$. Then $||a, n_i|| < ||a, a_M||$ for $\alpha \in [0, \frac{\pi}{3}[$.*

Proof. First we imagine a point a' that lies on the line from a_M to a with the same distance to a_M as n_i has (see figure 2.6(b)). The law of cosine states:

$$||a', n_i||^2 = ||a_M, n_i||^2 + ||a_M, a'||^2 - 2 \cdot ||a_M, n_i|| \cdot ||a_M, a'|| \cdot \cos(\beta)$$

Because of the construction of point a', it holds that

$$||a', n_i||^2 = 2 \cdot ||a_M, n_i||^2 - 2 \cdot ||a_M, n_i||^2 \cdot \cos(\beta)$$

therefore

$$||a', n_i|| = \sqrt{2 - 2 \cdot \cos(\beta)} \cdot ||a_M, n_i||$$

$$\Leftrightarrow ||a', n_i|| = 2 \cdot \sqrt{\frac{1 - \cos(\beta)}{2}} \cdot ||a_M, n_i||$$

We use a trigonometric transformation and obtain

$$||a', n_i|| = 2 \cdot \sin\left(\frac{\beta}{2}\right) \cdot ||a_M, n_i|| \tag{2.1}$$

Applying the triangle inequality we know that

$$||a, n_i|| \leq ||a', n_i|| + ||a, a'|| \tag{2.2}$$

and, of course, it holds that

$$||a, a'|| = ||a_M, a|| - ||a_M, a'|| = ||a_M, a|| - ||a_M, n_i|| \tag{2.3}$$

So, if we combine (2.1), (2.2) and (2.3):

$$||a, n_i|| \leq 2 \cdot \sin\left(\frac{\beta}{2}\right) \cdot ||a_M, n_i|| + ||a_M, a|| - ||a_M, n_i||$$

$$= \left(2 \cdot \sin\left(\frac{\beta}{2}\right) - 1\right) \cdot ||a_M, n_i|| + ||a_M, a||$$

$$= ||a_M, a|| - \underbrace{\left(1 - 2 \cdot \sin\left(\frac{\beta}{2}\right)\right) \cdot ||a_M, n_i||}_{=\Delta}$$

For $\beta \in [0, \frac{\pi}{3}]$ Δ is always ≥ 0, therefore $||a, n_i|| \leq ||a_M, a||$.
For $\beta \in [0, \frac{\pi}{3}[$ Δ is always > 0, therefore $||a, n_i|| < ||a_M, a||$. $\qquad\square$

Lemma 3. *There can be at most 6 sectors with one agent that chooses a_M as communication partner.*

Proof. Assume we have $s > 6$ sectors with an agent[8] n_i each. There is at least one pair of agents (n_{i_1}, n_{i_2}) with an angle γ between the lines from a_M to each of it with $\gamma < \frac{2\pi}{6}$. In Lemma 2 we have proven that for such a setting only one of these agents can choose a_M as a communication partner. This is a contradiction to our assumption, hence $s \leq 6$ follows.   □

Now we have all the information we need to prove the superordinated Theorem:

Proof. If we choose a number of sectors which is equal to 6, we can define β to be an angle $< \frac{\pi}{3}$ because of the definition that points on the border between two sectors belong only to the rightmost one and $\beta \leq \alpha < \frac{\pi}{3}$ (due to construction). Thus we know that for all agents $a \in A_i$ the agent n_i is located closer than the agent a_M (see Lemma 2). The in-degree can no more be increased, its maximum value is 6.   □

In the next Theorem we will expand this result for arbitrary k's, the proof is very similar to the one for Theorem 2.

Theorem 3. *In a two-dimensional space the maximum number of incoming connections to an arbitrary agent $a_M \in \mathcal{A}$ in a k-neighbourhood graph is $6 \cdot k$.*

Proof. Again we divide the space around agent a_M into 6 uniform sectors $S_1, ..., S_6$. Without loss of generality we consider one of these sectors denoted by S_i. Now we consider the k nearest agents $n_{i_1}, ..., n_{i_k}$ with $\delta(a_M, n_{i_1}) \leq \delta(a_M, n_{i_2}) \leq ... \leq \delta(a_M, n_{i_k})$ in each sector to the agent a_M (see figure 2.7). Let A_i contain the remain-

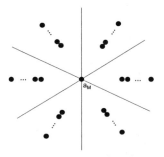

Fig. 2.7. The order of all nearest agents n_{x_y} in a way that they can choose a_M as one of the k nearest agents.

ing agents in this sector $A_i = \{a_j | a_j$ located in $S_i, a_j \neq n_{i_l} \forall l \in \{1, ..., k\}\}$. If each of these agents n_{i_x} can be located in the sector in such a way that the agent a_M is among the k nearest ones, the agent a_M has an in-degree of $6 \cdot k$. In figure 2.7 one can

---

[8] This is again the agent with the least distance to a_M in sector i

find an example how an in-degree of $6 \cdot k$ can be realised, for construction we use the same idea as we did in Lemma 1.

Now we can argue as we did in Theorem 2. If we consider any agent $a \in A_i$, the k nearest agents are among other agents in A_i (or maybe other agents in other sectors), or at least the k agents n_{i_k}. The reason is the same as in the proof for Theorem 2. There is no chance to choose the agent a_M as the nearest one, the n_{i_x} agents are always closer. Therefore, the in-degree of the agent a_M can never become higher than $6 \cdot k$. $\square$

2.3.4 kNG in Three-Dimensional Space

In the previous section, we presented a maximum number for the in-degree of agents connected by a k-neighbourhood graph. We gave an example how the maximum in-degree of $6 \cdot k$ could be realised. But these results were only valid for agents in two-dimensional space. In this section, we will concentrate on a three-dimensional space and will determine a lower and an upper bound for the in-degree. Therefore, we start by giving an example how an in-degree of $12 \cdot k$ can be constructed and will then prove that an in-degree greater than $60 \cdot k$ cannot occur.

Theorem 4. *In a three-dimensional space the in-degree for an arbitrary agent a_M can be equal to $12 \cdot k$.*

Proof. For this lower bound proof we construct an icosahedron $\mathcal{I}$ (see figure 2.8(a)) with side length l, vertex set $E = \{e_1...e_{12}\}$, and a_M as the centre. The outer sphere perimeter radius of $\mathcal{I}$ is

$$R = \frac{1}{4}\sqrt{10 + 2\sqrt{5}} \cdot l \approx 0.95106 \cdot l < l$$

We distribute $12 \cdot k$ agents in such a way that on each vertex there are exactly k agents located in a small ϵ environment. Next we show that, for any k, such an agent positioning results in an in-degree of exactly $12 \cdot k$ for agent a_M.

For $k = 1$, the analysis is very simple. If we locate agents on all vertices of $\mathcal{I}$ (i.e. $\epsilon = 0$), the agent a_M in the middle will have in-degree of 12 because for each agent, a_M is the nearest adjacent agent. This is due to the geometrical properties of an icosahedron concerning the side length and the inner sphere radius. Each vertex has a distance of at most R to the centre point, sometimes less. At the same time, the side length, i.e. the distance between adjacent vertices, is l and this is greater than R. Hence, the agent a_M in the centre is closer than any agent on the icosahedron vertices. Therefore, a_M is chosen by all 12 agents and has an in-degree of $12 \cdot 1$.

This idea can be extended for arbitrary k. If we consider $k > 1$, we can locate k agents in an $\epsilon \in \mathbb{R}_0^+$ environment around each vertex e_x. This environment can be seen as a sphere with radius ϵ and e_x in the centre. We denote the agents around an arbitrary vertex e_i by $A_i = \{a_1, ..., a_k\}$. By definition, the distance from vertex e_i to any $a \in A_i$ is bounded by ϵ (i.e. $\delta(e_i, a) \le \epsilon \ \forall a \in \mathcal{A}_i$). If we now choose

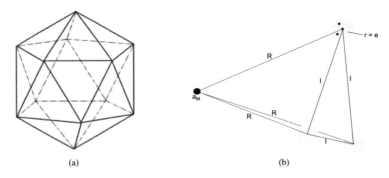

(a)   (b)

Fig. 2.8. Figure 2.8(a) visualises the geometrical shape of an icosahedron. It consists of 20 triangles of the same size and 12 vertices. In figure 2.8(b), one of these regular triangles and the ϵ-environment, where the agents can be located, is schematically displayed.

$$\epsilon < \frac{1}{3} \cdot \left(l - \frac{1}{4}\sqrt{10 + 2\sqrt{5}} \cdot l\right) \Leftrightarrow R < l - 3 \cdot \epsilon$$

and examine the icosahedron with side length l, it holds for each agent $a \in A_i$ that $(k-1)$ agents are in the same ϵ-neighbourhood (the agent will not choose itself as a neighbour). Due to the location of the ϵ-neighbourhood sphere we know that

$$||a, a_M|| \leq (R + \epsilon) \ \forall a \in A_i$$

At the same time, the distance between any agent $a \in A_i$ and an arbitrary agent $b \in A_j$ in the ϵ-environment of edge e_j $(i \neq j)$ is:

$$||a, b|| \geq (l - 2 \cdot \epsilon)$$

Therefore, the following inequality holds:

$$||a, a_M|| \leq (R + \epsilon) < (l - 3 \cdot \epsilon) + \epsilon = (l - 2 \cdot \epsilon) \leq ||a, b||$$

Hence, each agent in an arbitrary ϵ-environment will choose the remaining $k-1$ agents and then the agent a_M as communication partners. Thus, a_M has in-coming connections from all $12 \cdot k$ agents.   □

Definition 8 (3-face Icosahedron (3F-Icosahedron)). *We introduce a new geometrical figure which is similar to the icosahedron (see figure 2.8(a)). In the 3F-Icosahedron, we divide each of the 20 faces of the icosahedron with side-length l into three faces of equal size. Therefore, we consider a single face with the vertices A, B and C. These three vertices form an equilateral triangle (with the side length l). The median lines from each side cross in a unique centroid point S (see figure 2.9). Now we can construct three faces with the same area as shown in figure 2.9. This*

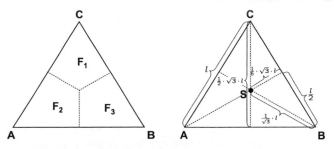

Fig. 2.9. These figures show the construction of the single faces of the 3F-Icosahedron. The geometrical properties of these faces are presented. The side length of the whole 3F-Icosahedron is l.

quadrangle has the side length $\frac{1}{6} \cdot \sqrt{3} \cdot l$ and $\frac{l}{2}$, the diagonals have the length $\frac{1}{\sqrt{3}} \cdot l$ and $\frac{l}{2}$. The resulting geometric object has obviously $20 \cdot 3 = 60$ faces. An example is shown in figure 2.10. From the geometrical properties of the icosahedron we know the in-sphere radius r

$$r = \frac{1}{12} \cdot \sqrt{3} \cdot \left(3 + \sqrt{5}\right) \cdot l \approx 0.755761 \cdot l$$

and the circum-sphere radius R:

$$R = \frac{1}{4}\sqrt{10 + 2\sqrt{5}} \cdot l$$

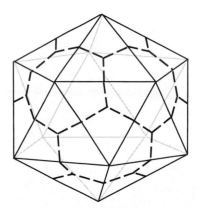

Fig. 2.10. The 3F-Icosahedron

Theorem 5. *In a three-dimensional space the maximum number of incoming connections to an arbitrary agent $a_M \in \mathcal{A}$ in a 1-neighbourhood graph cannot be greater than $60 \cdot 1$. This is an upper bound.*

Proof. We start with the construction of a 3F-Icosahedron $\mathcal{I}$ as defined before. The side length of $\mathcal{I}$ is variable and denoted by l. The centre of $\mathcal{I}$ is the agent a_M. The whole space around a_M can now be divided into 60 pyramidical structures $\mathcal{S} = \{S_1, ..., S_{60}\}$. Each S_i can be seen as the extension of a pyramid, with a_M as apex and one face of $\mathcal{I}$ as basis. Without loss of generality, we consider an arbitrary segment $S \in \mathcal{S}$.

In S, we consider the agent $a \in \mathcal{A}$ with the shortest distance to a_M. We modify now the side length of $\mathcal{I}$ in such a way that a lies directly in a face of $\mathcal{I}$. An example is shown in figure 2.11. Now we will prove that the distance from an arbitrary agent b

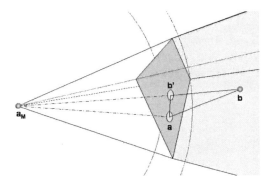

Fig. 2.11. Schematic figure of the proof for Theorem 5

that lies in S to this agent a is always smaller than the distance from b to agent a_M. Hence, the in-degree of a_M is bounded by the number of faces of the icosahedron. Therefore, we have to show: $||b, a|| < ||b, a_M||$.

We know that all points on the surface of a 3F-Icosahedron face lie between or on the in- and the circum-sphere, therefore we know that

$$r \leq ||a_M, a|| \leq R \qquad (2.4)$$

We can add the projection[9] of b (denoted by b') onto the face into this relation, then we can restate our goal to

$$||a_M, b'|| + ||b', b|| = ||b, a_M|| > ||b, a||$$

If we add here equation (2.4), we can consider the minimum value for $||a_M, b'||$ and replace this part by its minimum value r, hence it is enough to show that

$$r + ||b', b|| > ||b, a|| \qquad (2.5)$$

---

[9] The projection of b onto an icosahedron face is the intersection point of the connection between b and a_M and the face itself.

Because the triangle inequality holds, we know that

$$||b, a|| \leq ||b', b|| + ||b', a|| \Leftrightarrow ||b', b|| \geq ||b, a|| - ||b', a|| \tag{2.6}$$

If we combine (2.5) and (2.6), we get

$$r + ||b, a|| - ||b', a|| > ||b, a|| \Leftrightarrow r > ||b', a||$$

We know from our definition of the 3F-Icosahedron that the longest distance in one face is

$$d = \frac{1}{\sqrt{3}} \cdot l \approx 0.57735 \cdot l$$

Therefore, the distance from b' to a is bounded by d and we can say that

$$r = \frac{1}{12} \cdot \sqrt{3} \cdot \left(3 + \sqrt{5}\right) \cdot l > \frac{1}{\sqrt{3}} \cdot l = d$$

This statement is true for arbitrary $l > 0$, thus we have shown that in each sector only the agent which is closest to a_M can choose this agent as communication partner. In each cone the agent a is some kind of barrier for further connections of other agents with a_M. Hence, we have proven an upper bound of $60 \cdot 1$ for the number of incoming connections. $\square$

Now, we will prove the upper bound for incoming connections for arbitrary k.

Theorem 6. *In a three-dimensional space the maximum number of incoming connections to an arbitrary agent $a_M \in \mathcal{A}$ in a k-neighbourhood graph cannot be greater than $60 \cdot k$.*

Proof. Assume, we have more than $60 \cdot k$ agents with connections to the agent a_M. If we again construct an 3F-Icosahedron around a_M, then there has to be at least one cone-like structure with a_M as apex (cf. figure 2.11) with $k' > k$ agents that have connections to a_M. Now consider the $(k + 1)$-nearest agent to a_M. With the same argumentation as we did in the proof for Theorem 5 we can show that this agent will choose as its communication partners each of the k-nearest agents to a_M prior. Hence, there is no possibility left to choose a_M as communication partner and this is a contradiction to our assumption. $\square$

If we now combine the results from Theorem 4 and Theorem 6, we have a lower and an upper bound for the maximum number of communication partners for arbitrary agents in a kNG. This number $i \in [(12 \cdot k); (60 \cdot k)]$ is independent from the overall number of agents.

2.3.5 Graph Connectivity

In this section we want to examine reasonable values for the size of a neighbourhood. Since large k values for the neighbourhood will offer a more complete communication between all nodes or agents and give access to more information, but produce higher costs at the same time, we are interested in a k value as small as possible to cope with this trade-off. When we consider the last section, this becomes even more urgent since an increase of k by one will sixfold (or up to sixtyfold for the three-dimensional case) the number of incoming communication lines in the worst case. Hence, we define a useful k to be the minimal value that creates a connected graph for all agents. In [KMN06], Kozakova et al. have proven that the size of the clusters is decreasing superexponentially when the dimension for the points will be increased. But we are interested in the connectivity of the agent set for a given dimension and varying values of k. Hence, we show some results for the development of the graph connectivity in the two-dimensional space.

First, it is quite obvious that, for worst case scenarios, k has to be at least half of the number of agents. This can be seen in the following Lemma.

Theorem 7. *If we consider n agents in an Euclidean space, we need at least a k-neighbourhood graph with $k = \lceil \frac{n}{2} \rceil$ to assure that the resulting graph is connected.*

Proof. We split the proof into two parts, first we show, that there exist settings such that a kNG with $k < \lceil \frac{n}{2} \rceil$ is to small to guarantee a connected graph, then we show that $k = \lceil \frac{n}{2} \rceil$ is large enough to connect all possible agent settings.

The first part will be done by presenting a counter-example. Lets consider a kNG with $k = 7 < 8 = \lceil \frac{n}{2} \rceil$ in the setting presented in figure 2.12. If it holds that $D > d$,

Fig. 2.12. Counter-example for Theorem 7

all agents in the left area will only create connections to the seven other agents that have a maximum distance of $d < D$. The same holds for all agents in the right area. Hence, there will be no connection drawn from any agent in the left area to any agent in the right area and the graph is not connected.

The second part of the proof is as simple as the first. Assume the graph $G = (V, E)$ is not connected for $k = \lceil \frac{n}{2} \rceil$. Then we have l subgraphs $G_1...G_l$ with no edge (i, j) between arbitrary two subgraphs, or, to put it in a more formal way:

$$\neg \exists (i, j) \in E | i \in G_o, j \in G_p \; \forall o, p \; (o \neq p) \in \{1, ..., l\}$$

We consider the graph $G_{l'}$ with the minimal number of nodes $|G_{l'}| = m$. m can only have values between 1 and $\lfloor \frac{n}{2} \rfloor$. Now we consider an arbitrary agent from the subgraph $G_{l'}$. This agent has $k = \lceil \frac{n}{2} \rceil$ outgoing connections to other agents, but in $G_{l'}$ there are at most $\left(\lfloor \frac{n}{2} \rfloor - 1 \right)$ other agents left, therefore it will connect to another agent in another subgraph. This violates our assumption. $\square$

Unfortunately, we just have proven a dependency between the neighbourhood size and the number of agents. But this dependency is only important for special cases, therefore we have investigated sufficient k values for random settings. Eppstein et al. present in [EPY97] a theoretical approach and have proven that the expected number of components in a graph where each node is connected with its nearest neighbour (it is the same as our 1-neighbourhood graph) is approximately $0.31 \times |V|$. This is the result we obtain with our simulation runs. In these experiments, we distribute n agents randomly in an Euclidean space and then connect the agents by a kNG. We count the number of disconnected graphs (clusters) for different values of k. In figure 2.13, the average and highest number of clusters are visualised for $k \in [1; 15]$. We

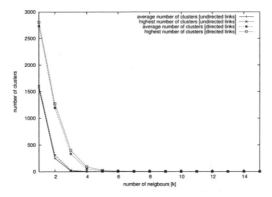

Fig. 2.13. The figure shows the number of clusters for different k-values. We considered 1000 random settings with 5000 agents each. The average and the highest number of clusters that occurred are visible.

distinguish between graphs with directed and undirected edges. Next, we change the number of agents to verify if good values for k in random settings are appropriate for any quantity of agents. The results can be found in figure 2.15. To summarise, the development for all number of agents in the simulation was similar, the number of clusters is decreasing very fast and simultaneously the average size of the largest clusters rises very fast to approximately 100% (see figure 2.16). For k values above 10, there is a probability very close to 1.0 to obtain a totally connected graph. And figure 2.14 shows that in graphs the theoretical maximum factor of $6 \cdot k$ factor can be adjusted to at most $2 \cdot k$ if random scenarios are considered.

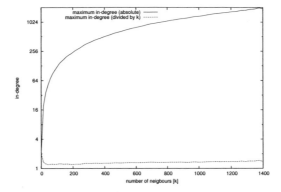

Fig. 2.14. The maximum in-degree for different k-values is presented in this figure. The y-axis is set to logarithmic scale. If this value is divided by k, we can see that in random settings the maximum in-degree is always below $3 \cdot k$ and mostly significantly below $2 \cdot k$.

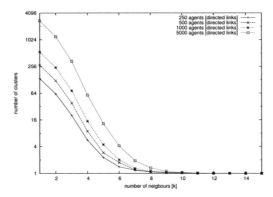

Fig. 2.15. This figure shows that for higher number of agents there is a slightly higher number of clusters, especially for small sized neighbourhoods. We changed the y-axis scale to logarithmic to increase the readability.

2.3.6 Energy Efficiency

In this section we will present an empirical analysis of our graph structure in comparison with another frequently used graph. We will compare the necessary energy (communication costs) to establish a connected graph for both graph types.

Communication Costs in Radial Graphs

We start with a graph structure which is common in the AdHoc network research. Here, each node has a fixed communication radius and communication with other nodes is only possible inside this radius. In figure 2.17 we present the correlation

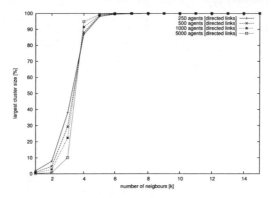

Fig. 2.16. This figure visualises the development of the size of the largest connected agent cluster or subgraph. For instance, a value of 80% means that the largest cluster contains 80% of all agents on average.

between communication radius and the number of disconnected clusters. We tested for each radius between 0% and 100% (in relation to the simulation space size) how many clusters appear in the resulting graph. The presented values are mean values of 100 runs which means that we count the number of clusters in 100 random settings for each radius. One settings contains 1000 agents. We conclude from this figure that

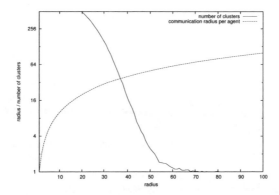

Fig. 2.17. Radial graph communication costs. The radius is the percentage of the area covered by a circle with the given radius in relation to the size of the whole space. The communication radius per agent is proportional to the value of the radius.

we need a radius greater than 70% of the whole space in order to obtain in nearly all settings a connected graph. At the same time, we define the communication costs linear dependent from the radius. Hence, each agent produces cost greater than 70 to achieve a totally connected graph.

Communication Costs in kNG

We use the same settings and parameters for the cost examination when using k-neighbourhood graphs. In figure 2.18, we present the results for different values of k. In these experiments, a k value greater than 5 creates a totally connected graph. At the same time, in k-neighbourhood graphs the average radius for each agent is only approximately 43, the maximum is at around 78. Again, these values are the average results from 100 runs[10].

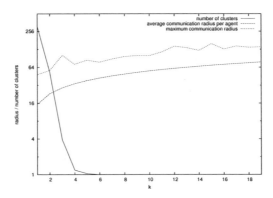

Fig. 2.18. kNG communication costs. For a given k, the average and maximum communication costs are visualised.

Comparison

In the preceding sections we showed the costs that were necessary to create a connected graph in our simulations. For these settings, the relation of costs between our graph and the radial graph is $\frac{43}{70} \approx 0.6$, hence we can save around 40% of the energy needed for communication. At the same time, the maximum radius in the kNG is with 78 only slightly higher than the radial graph communication radius. If we consider communication costs closer to real hardware systems, the energy consumption increases quadratic or even more with the distance (see for example section 8.1 for the distance influence on data rate using wireless LAN). In such a setting, we can save more than 60% of the energy ($\frac{43^2}{70^2} \approx 0.38$). But such a consideration of real hardware systems shows on the other hand that a modulation of the communication range is not as simple as we wish it to be, and we do not consider the communication cost for the initial buildup of our network structure. When we deal with static and persistent networks and / or the distance between the nodes is very large, our approach has a clear potential to save energy.

---

[10] Except the value for the maximum radius, this is the maximum value over all runs.

2.3.7 Conclusion

In the latter part of this chapter we were engaged with a special type of graph, the k-neighbourhood graph. We have geometrically proven that in such graphs with k outgoing edges to the nearest agents the in-degree of each agent is bounded by $6 \cdot k$. In three-dimensional space, we could prove a lower and an upper bound of $12 \cdot k$ and $60 \cdot k$ with the construction of a special geometrical structure called 3F-Icosahedron. Additionally, we showed with experiments that - in spite of a very bad worst case behaviour - in random settings very small values for k are sufficient to generate a connected graph between all agents. With a kNG, locally obtained information can be spread globally, independent from the overall number of agents. Therefore, this communication graph structure seems to fit perfectly into the swarm intelligence area. In further experiments, we compared the kNG communication structure with the classical and often used radial communication structure concerning energy costs. In experimental settings we found out that we can save a lot of energy (more than 40%) and still can have a connected graph even if we are using only small values for the parameter k.

3

The Online Partitioning Problem

The problem is not that there are problems.
The problem is expecting otherwise and thinking that
having problems is a problem.

Theodore I. Rubin

In this chapter we introduce the optimisation problem examined in this thesis. It will first be formally defined and we will present an algorithm with exponential runtime that solves arbitrary *OPP* instances. For two targets ($|T| = 2$), an algorithm with runtime $O(n^2)$ which can easily be reduced to $O(n \cdot \log n)$ is presented and its correctness is proven.

In the next large part, related areas and research in computer science dealing with problems similar to the *OPP* are listed. We only give a rough idea of these topics. Thus, similarities and differences can be understand.

In the last section, a formal method to describe and model multi agent systems is presented. The agents and the objectives of the *OPP* could be fitted in the single elements of the model.

3.1 The OPP

In this work, we consider a multi-objective optimisation problem, the *Online Partitioning Problem (OPP)*, for the coordination of large groups of agents. The idea behind this problem is that we want to associate agents to tasks. Both agents and tasks are located in a Euclidean space of unknown dimensions. Therefore, we refer to tasks as targets.

The *OPP* takes several objectives into account. First, the association of the agents with targets should be uniform, i.e. if we have n agents and k targets in a good solution the number of agents associated with each target should be bounded by $\lceil \frac{n}{k} \rceil$. Second, the distance from an agent to an associated target is considered. For an optimal solution, the overall distance between the agents and the targets should be minimal. The third objective is quite imprecise. We want the abilities of the agents to be as limited as possible. We do not give this formal expression, but the idea behind is that each additional ability for every agent will increase the costs for such a system significantly. Hence, we do not want a central server that has the ability to coordinate the whole group. As a result we have the following objectives:

(1) The agents have to be distributed uniformly (distribution).
(2) The overall distance has to be minimised (distance).
(3) The abilities of the agents should be very simple (ability).

These three objectives are in mutual opposition, so we look for the best possible solution fitting in all objectives in a quite acceptable way.

We can put these objectives in a more formal way by defining the first two objectives mathematically

Definition 9. *An instance of the OPP with $OPP = (\mathcal{A}, \mathcal{T}, \rho)$ consisting of the agent set $\mathcal{A} = \{a_1, ..., a_n\}$, the target set $\mathcal{T} = \{T_1, ..., T_m\}$, and a position function ρ is an instance to an optimisation problem. The question is whether there exists a partition $\{S_1, S_2, ..., S_m\}$ of $\mathcal{A}$ with $S_i \subseteq \mathcal{A}$ $(i \in \{1, ..., m\})$ such that $\mathcal{A} = \{S_1 \cup S_2 \cup ... \cup S_m\}$, which simultaneously maximises*

$$\prod_{i=1}^{m} |S_i|$$

and minimises

$$\sum_{i=1}^{m} \sum_{a \in S_i} \delta(a, T_i)$$

This can be combined in the following equation[1]. There, b_i denotes the number of agents that have chosen the target T_i in the current partitioning decision and o_i the number of agents that would have chosen target T_i in an optimal partitioning. $\tau(a_i)$ denotes the target currently chosen by agent a_i.

$$f = \alpha \cdot \left(\frac{\prod_{i=1}^{m} b_i}{\prod_{i=1}^{m} o_i} \right) + \beta \cdot \left(\frac{\sum_{i=1}^{n} \min_{j=1..m} (\delta(a_i, T_j))}{\sum_{i=1}^{n} \delta(a_i, \tau(a_i))} \right) \tag{3.1}$$

under the constraints $\alpha + \beta = 1$ and $\alpha, \beta \geq 0$ for the weights α and β.
The higher the value f ($f \in [0; 1]$), the better is the solution found.
For this mathematical description we use the fact that a product of m numbers is maximal if the numbers are identical.

---

[1] The set of optimal solutions build a pareto front. In the remaining part of the thesis, we combine both objectives to one function weighted by two factors.

The proof can be found in the appendix (see appendix A.1).

For the third objective, we consider the effects of different abilities to the solution quality of the problem described with equation 3.1. In most of the advanced approaches we utilise inter-agent communication to enhance the knowledge and therefore the basis for appropriate decisions. This communication or information sharing can generate costs in the form of additional hardware requirements for each agent and in form of energy costs for the transmission of these information. For this purpose we incorporate these costs with the third objective. Hence, equation (3.1) is mostly only a part of the whole evaluation of the approach.

The knowledge of the agents is limited to the local neighbourhood around them and (in most approaches) to some - relative or absolute - information about the distances to the existing targets. As in most other approaches, the agents are infinitely small and they cannot collide with each other.

3.2 Exact *OPP* Algorithms

In the previous section we present a definition for the problem with which the remaining part of the thesis deals. Next, an exact algorithm is presented that can solve an *OPP* instance for arbitrary number of agents and targets exactly. Even though this algorithm is acting in a global manner, it still has an exponential runtime ($O(m^n)$). Only for instances with two targets ($m = 2$) we could find an algorithm which is much faster, in detail it is in $O(n^2)$. This motivates the development of heuristics the remaining chapters of this thesis deal with.

The abilities of the agents are fixed in these exact analyses. Hence, we do not regard the third objective.

Algorithm 3 A complete algorithm for the *OPP* which needs $O(k^n)$ runtime.

```
 1: procedure CALCULATEOPPEXACT(AgentList agents, int n, int m)
 2:     if (n == 0) then
 3:         currentSolution ← calculateOPPSolutionQuality(agentSet)
 4:         if (currentSolution > optimalSolutionQuality) then
 5:             optimalSolutionQuality ← currentSolution
 6:             optimalAssignedAgents ← agentSet
 7:         end if
 8:     else
 9:         for (i = 1; i ≤ m; i + +) do
10:             τ(aₙ) ← i
11:             calculateOPPExact(agents, (n − 1), m)
12:         end for
13:     end if
14: end procedure
```

3.2.1 Complete Algorithm for Arbitrary Parameters

Starting with an exact algorithm for arbitrary numbers of agents and targets, we can offer an exact solution for any *OPP* instance. Algorithm 3 is proceeding as follows: Each possible combination of assignments of the agents to the targets will be iterated in a recursive algorithm. For every intermediate solution which assigns all of the n agents to a specific target, the quality of this solution will be determined by the quality function introduced in equation 3.1. Both the best solution found so far and the associated assignment is stored in a global variable. After completion of the algorithm, these global variables contain the optimal solution.

Heuristic with Global Knowledge

All approaches in this thesis use limited agent knowledge and no central instance. With such a central instance, a very simple heuristic could be developed. An example is shown in Algorithm 4. It is a greedy approach that assigns in each round to each

---

Algorithm 4 A heuristic for the *OPP* which uses a central instance.

1: **procedure** CALCULATEOPPSOLUTIONWITHGLOBALKNOWLEDGE(AgentList agents, int n, int m)
2:    **for** $(i = 1; i \leq m; i + +)$ **do**
3:       $agentList_i \leftarrow sort(agents, T_m)$ ▷ sort agents according to target position
4:    **end for**
5:    **repeat**
6:       **for** $(i = 1; i \leq m; i + +)$ **do**
7:          Agent a ← nearest, not assigned agent in $agentList_i$
8:          assign a to target T_i
9:       **end for**
10:   **until** (all agents assigned to one target)
11: **end procedure**

---

target the nearest - currently unassigned - agent. It runtime is in the order of $O(m \cdot n \cdot \log n + m \cdot n)$ or $O(m \cdot n \cdot (1 + \log n))$ due to the sorting procedures. We use this algorithm whenever the runtime of the complete algorithm is too high.

3.2.2 Exact Algorithm for Two Targets

In the last section we presented an algorithm with runtime $O(m^n)$ which is extremely bad for large numbers of n. In this section, we will present an exact algorithm for two targets that runs in $O(n^2)$ which can be easily reduced to $O(n \cdot \log n)$.

The idea of this algorithm is as follows. We sort the agents according to the distance difference

$$d_i = \delta(a_i, T_1) - \delta(a_i, T_2) \quad \forall i \in \{1, ..., n\}$$

The agents with non-negative distance difference will be assigned to target T_2, the others to target T_1.

Then we sort the agents in ascending order according to these differences in a list $\mathcal{L}$ in such a way that

$$\mathcal{L} = a_{k_1}, a_{k_2}, ..., a_{k_n} \text{ with } d_{k_1} \leq d_{k_2} \leq ... \leq d_{k_n}$$

Let $part_{T_x}(\mathcal{L})$ denote the number of agents that are assigned to target T_x ($x \in \{1, 2\}$). Now we consider the agent $a_{k_j} \in \mathcal{L}$ having the smallest positive or the biggest negative value. We choose the smallest positive value if the number of agents currently assigned to target T_2 is greater than or equal to the number of agents currently assigned to target T_1. This can be expressed by

$$j = \begin{cases} part_{T_1}(\mathcal{L}) \text{ for } part_{T_1}(\mathcal{L}) > part_{T_2}(\mathcal{L}) \\ part_{T_1}(\mathcal{L}) + 1 \text{ for } part_{T_2}(\mathcal{L}) \geq part_{T_1}(\mathcal{L}) \end{cases}$$

Without loss of generality, we assume for the further steps that there are more agents assigned to target T_2.

The agent a_{k_j} chosen by the previous equation will now be reassigned from target T_2 to target T_1. After this, we calculate the solution quality again. If the quality could be enhanced, we continue with our algorithm and search for the next a_{k_j}. This algorithm is presented in pseudocode in Algorithm 5. It has a runtime that lies in $O(n^2)$, but it can easily be modified to a runtime of $O(n \cdot \log n)$ by using some more intelligent data structures (for example *Heaps*). Then, the most time consuming part is the sorting procedure in line 10; the FOR and the $REPEAT$ loop can both run in time $O(n)$ with some small constants.

Proof of Correctness

In this section, we will prove that the algorithm is able to find an optimal solution for all possible *OPP* instances with two targets. The output of Algorithm 5 is always a list of agents, starting with the agents assigned to target T_1, followed by agents assigned to T_2. We start with a proof that there is always at least one optimal solution for a given setting that has such a form. Then, we will show that our algorithm will find this solution.

For the Theorems, we introduce some additional notations.

A solution $\mathcal{S} = (s_1, s_2, ..., s_n)$ for the *OPP* can be seen as a list of triples $s_i = (id(i), d_i, t_i)$ with $id(i)$ denoting a function that calculates the index of the ith agent in this solution, d_i standing for the distance differences $\delta(a_{id(i)}, T_1) - \delta(a_{id(i)}, T_2)$ and t_i reflecting the currently chosen target by the $id(i)$th agent.

Theorem 8. *For any optimal solutions $\mathcal{S} = (s_1, s_2, ..., s_n)$ for the OPP with $s_i = (id(i), d_i, t_i)$ there exists at least one solution $\mathcal{S}' = (s'_1, s'_2, ..., s'_n)$ having the same solution quality as the optimal one with the following property. For $\mathcal{S}'$ it holds that $s'_i = (id'(i), d'_i, t'_i)$ and $d'_i \leq d'_{i+1}$ in such a way that there is exactly one split position $j \in \{1, ..., n\}$ with $T_1 = t'_1 = t'_2 = ... = t'_j \neq t'_{j+1} = t'_{j+2} = ... = t'_n = T_2$*

Algorithm 5 An optimal algorithm for the *OPP* with two targets that needs a runtime of $O(n^2)$.

1: **procedure** CALCULATEOPPSOLUTION(AgentSet agentSet, Target T_1, Target T_2)
2:    **for all** (agents a in agentSet) **do**
3:        $a.calculateDifference(T_1, T_2)$
4:        **if** $(a.getDifference \geq 0)$ **then**
5:            $a.setTarget(T_2)$
6:        **else**
7:            $a.setTarget(T_1)$
8:        **end if**
9:    **end for**
10:   List L $\leftarrow$ $agentSet.sortAscendingAccordingToDifferences()$
11:   $finished \leftarrow FALSE$
12:   **repeat**
13:       $agentSetOld \leftarrow agentSet$
14:       **if** $(part_{T_1}(L) > part_{T_2}(L))$ **then**
15:           $a \leftarrow L.getAgentOnListPosition(part_{T_1}(L))$
16:           Agent $a.setTarget(T_2)$
17:       **else**
18:           $a \leftarrow L.getAgentOnListPosition(part_{T_1}(L) + 1)$
19:           Agent $a.setTarget(T_1)$
20:       **end if**
21:       **if** $(agentSet.getSolutionQual() \leq agentSetOld.getSolutionQual())$ **then**
22:           $agentSet \leftarrow agentSetOld$
23:           $finished \leftarrow TRUE$
24:       **end if**
25:   **until** (finished == TRUE)
26: **end procedure**

Proof. Assume, there is an optimal solution $\mathcal{S} = (s_1, s_2, ..., s_n)$ with fitness f_{opt} calculated by combining the partitioning fitness $f_p(\mathcal{S})$ and the distance fitness $f_d(\mathcal{S})$. We assume not to have any solution $\mathcal{S}' = (s'_1, s'_2, ..., s'_n)$ with only one split position. Let s'_k be the last occurrence of target T_1 in the list and s'_l the first occurrence of T_2 such that for k, l the following equation is true:

$$\forall s'_i \in \mathcal{S}' : ((t'_i = T_1) \wedge (k \geq i)) \vee ((t'_i = T_2) \wedge (l \leq i))$$

If T_1 (T_2) does not occur in the list, the appropriate value is n (0). Because of our assumption, we know that $k > l$. If we now swap the chosen targets of s'_k and s'_l, the partitioning quality of the solution stays the same regarding to this aspect, hence $f_p(\mathcal{S}) = f_p(\mathcal{S}')$. Considering the distance part $f_d(\mathcal{S})$ of the fitness, there are the following modifications: The overall distance value will be reduced by preliminary decisions of the kth and lth agents and increased by the new distances to the other targets, respectively.

$$f_d(\mathcal{S}') = f_d(\mathcal{S}) - \delta(a_k, T_1) - \delta(a_l, T_2) + \delta(a_k, T_2) + \delta(a_l, T_1)$$

This can be written as

$$f_d(\mathcal{S}') = f_d(\mathcal{S}) + (\delta(a_l, T_1) - \delta(a_l, T_2)) - (\delta(a_k, T_1) - \delta(a_k, T_2))$$

And now we can imply the distance difference values

$$f_d(S') = f_d(S) + d'_l - d'_k$$

If $d'_k \neq d'_l$: Since $d'_k \leq 0$ and $d'_l \geq 0$, $f_d(S)$ will be definitely increased. This contradicts our assumption that $f_d(S)$ is an optimal solution.

If $d'_k = d'_l$: The solution quality will not be changed, hence we can exchange the two target decisions. This step can be repeated as long as there are more than one split positions. After these steps, we have a solution with only one split position which is optimal. This is a contradiction to our assumption.

In either case, we have a contradiction. Hence, the assumption is false and the Theorem is proved. □

Theorem 9. *Algorithm 5 is able to find an optimal solution.*

Proof. Because of the previous Theorem we can say that an optimal solution $S = (s_1, s_2, ..., s_n)$ for an arbitrary setting has exactly one split point j with $T_1 = t_1 = t_2 = ... = t_j \neq t_{j+1} = t_{j+2} = ... = t_n = T_2$. If there are several solutions with one split point each, we focus on the smallest (with regard to the index j) one. The intermediate assignment during execution of the algorithm is denoted by $S' = (s'_1, s'_2, ..., s'_n)$.

We start by showing that the algorithm moves in the correct direction with its target exchanging. After that, we show that the fitness increase is strictly monotonic and ends when the optimal solution is reached.

Due to the initialisations of the agents to the nearest targets, there exists no better value for the distance part of the starting target assignment $f_d(S')$, hence we consider the distribution aspect f_p of the solution S'.

We can distinguish two cases. The first one is that $f_p(S') = f_p(S)$. In this case the algorithm found the optimal value and will terminate since any target exchange cannot improve the solution. Hence, we consider the situation that $f_p(S') \neq f_p(S)$. Due to the optimality of S, the overall fitness $f(S) > f(S')$, and $f_d(S) \leq f_d(S')$, $f_p(S')$ has to be smaller than $f_p(S)$. Hence, the search direction of the algorithm is correct.

Next, we will show the strictly monotonic increase of the fitness values during execution of the algorithm. Let S_i be the ith intermediate solution. The algorithm will change the target assignment of one element if it can gain a fitness increase, i.e. $f(S_i) > f(S_{i-1})$. We consider the last execution / step of the algorithm. There, it will change the target value of element s'_{j-1} to the value of element s'_j. We know that the fitness after the exchange is higher than the fitness before the exchange. We additionally know that the partitioning fitness increase is the same for all elements and has the value $2n^{-1}$. If in this step the fitness gained by enhancing the partitioning is bigger than the fitness lost by degrading the distances, the same must hold for all steps executed by the algorithm before, since the distance loss is increasing with every step. Hence, each step of the algorithm enhances the overall fitness until the optimum is reached. □

3.3 Related Problems

There are several problems that have similarities with the *OPP*. These will be presented in the following sections. Herewith, the *OPP* can be integrated in recent and classic research areas. At the same time, the differences can be explained and comprehended.

3.3.1 Pattern Formation

The pattern formation problem is widely defined as the coordination of groups of agents [BSS03]. These agents are basically located in the area of robotics and the problem is to form a certain shape, such as a circle, a chain, a wedge etc. We consider a formation as a group of mobile agents established in a predefined geometric shape. Such a shape should be maintained even when passing obstacles or adding / removing agents. Applications for pattern formation algorithms are for example search and rescue operations, remote terrain and space exploration, arrays of satellites or unmanned aerial vehicles (UAV) controlling. Moving in a formation can also limit the sensor costs for each single agent in the group [BH00]. In nature, several forms of pattern formation can be observed, too. One of the most famous is the bird flocking phenomenon, first described by Reynolds in [Rey87]. He developed a simple egocentric behavioural model for flocking which is instantiated in each member of the simulated group of birds. Other examples are fish schooling and several ant behaviours, for example the food chain formation [CDF+01].

> *From the robotic side, swarms self-organise into patterns. From the biological side they construct ordered patterns. [Ben04]*

The area of pattern formation can be divided into two main parts, the centralised and the distributed control. We will concentrate on the latter one since this is closer related to the scope of this thesis, but we will give a rough overview on the first one for completeness. This classification is based partly on a review of Bahceci et al. [BSS03].

Centralised Pattern Formation

Centralised pattern formation systems have one computational unit that acts as a global coordinator. It knows or can sense all other agents or robots, can compute a solution, and can transfer orders to each agent via a communication channel. The main problems in this area are to deal with inexact sensor information and to compute correct paths of quick moving agents fast enough. Examples for such centralised systems can be found for instance in [BK02, EH01, KS01].

Decentralised Pattern Formation

Systems with no centralised instance can again be divided into two fields. We can distinguish between systems where each robot has global information and systems where the information of each robot is limited. A nice overview of systems with global information has been made by Sugihara et al. [SS96] and Suzuki et al. [SY99]. In the latter, the class of geometric patterns that the robots can form in terms of their initial configuration is characterised and some impossibility results are also presented. But, as a disadvantage of these approaches, the agents have to be able to store all previous decisions, hence they are non-oblivious. [DK02] present an algorithm for the construction of a circle for oblivious agents. As in most other approaches, the agents are infinitely small and they cannot collide with each other. They show that synchronously acting agents can easily be captured in deadlock situations. In [FPSW99], Flocchini et al. analyse the possible patterns that can be realised and they try to find the minimal abilities of the agents to fulfil such patterns[2]. To give an example, [FM02] present an idea in which each agent orients itself to the nearest and the farthest agent. This information is extracted and deducted from the global knowledge. In all of these results, each robot is capable of recognising the instantaneous position of every other robot at any point in time. This is again very different from our approach since such a system is not very scalable and has nothing in common with biological systems.

Hence, we concentrate on research in the field of decentralised pattern formation algorithms for agents with limited visibility. In this area, Prencipe et al. [FPSW01] have made comprehensive studies. They present a theoretical examination of an algorithm to gather agents in an arbitrary point. Therefore, agents need a common coordinate system (i.e. a common sense of direction). This is necessary because the gathering problem is unsolvable if the robots are disoriented, which is proven in [FPSW01]. Another problem mentioned in the same paper is the unsolvability of the problem if the visibility graph of the agents is disconnected, i.e. if there are at least two groups of agents that cannot sense and see each other, respectively.

Fujibayashi et al. [FMSY02] present an idea where the agents are connected by virtual springs that maintain a desired formation during obstacle avoidance. Arbitrary formations could be reached by breaking springs with a predefined probability, but these probability values are hard to discover. [BSS03] also list several approaches that make use of graph-based approaches. All of them use control graphs to maintain and verify the demanded structure.

Balch et al. present in [BH00] an approach inspired by molecules forming crystals. The main focus in this approach lies on obstacle avoidance. The desired formation is a parameter given to the system at the beginning and the agents already start with correct positions in this formation. Balch examines how the group can have benefits from the establishment of the formation and how good the formation can be main-

---

[2] They place their work in contrast to [SY99] regarding the limitations of the agents. But the differences they make in the knowledge of agents is only the number of common axes with / without common knowledge on their direction.

tained while avoiding obstacles. The maximum number of agents they dealt with was 32.

Kostelnik et al. add local communication to the system in [KSJ02]. The agents have a common coordinate system. The authors propose a variation of a hierarchical model where only one leader is present and where each agent informs all hierarchical superior agents about its local sensible part of the formation. The leader has global knowledge and can decide which agents have to change position to establish the formation. Therefore, this system works with a central instance but this instance will be found in a self-organising process. Due to the limitations of the communication radius, multiple smaller occurrences of the desired formation can form themselves. Again, the paper deals with very small sets of agents, the biggest simulations contain 7 agents.

[FM02] also use local communication. The approach is quite similar, but the ideas are described in more detail and with more references to a real algorithm. The set of possible shapes is described by *those shapes that can be folded from an open bicycle chain, keeping either the middle or the end of the chain in front*. The idea of the algorithm is as follows:

All agents in a formation send continuously a signal through the whole formation. Hence, all agents know how many agents are in the formation and based on special shape descriptions each agent can arrange its position relative to another agents' position. The agents have some internal order based on their position in the formation. This algorithm has been successfully tested on up to 6 Pioneer2 DX robots.

3.3.2 Dynamic Task Allocation

The algorithms presented in this thesis can be used for dynamic task allocation, the control of groups of simple robots / weapons in military and civilian fields, or even for the division of a large group of humans without an explicit hierarchy structure onto some targets, tasks or streets etc., as done in a large experiment with conference attendees. [PKea03]

The main idea of Dynamic Task Allocation (DTA) is to find an assignment of tasks to machines. At the same time, the DTA is a decentralised problem. There is no global optimiser or scheduler that can calculate an optimal solution. We applied *OPP* algorithms to a DTA problem in chapter 8. A more detailed description of DTA problems can be found there.

3.3.3 Graph Partitioning

Graph Partitioning algorithms (GP) are a well known and extensively studied research area in computer science. There exists a large number of heuristics to cope with this NP-complete [GJS76] problem. Most of these approaches use the whole graph as the input of the algorithms or build the graph during the runtime of the calculation [SWC04, HL95, Jon92]. We distinguish three areas of graph partitioning ideas. First, classical GP algorithms will be briefly presented for graphs without coordinate information. Then, some ideas for graphs with coordinate information will be shown. These solutions are more closely connected to the problem we are dealing with. In the third part, multi agent system and / or biologically inspired solutions will be explained. Additionally, we describe some clustering ideas since this topic is connected to the area of graph partitioning.

In this section, we describe a Graph $G = (V, E)$ by a set of vertices and edges. Both edges and vertices can have an associated weight. The task of the algorithm is to split the graph into k disjoint subgraphs with approximately the same amount of vertex weight while minimising the sum of the weights of the edges whose incident vertices belong to different subgraphs [SKK00].

Partitioning Graphs without Coordinate Information

The partitioning algorithms can be divided into two parts, on the one hand statistical iterative approaches that obtain solutions in the most effective way but need the number of clusters as an input. On the other hand, clusterings can be composed by hierarchical algorithms.

One of the most famous partitioning or clustering approaches is the k-means algorithm. Starting from a random partitioning, cluster centres will be calculated. The nodes or data points will be assigned each to the closest of these centres. 'Closest' means in this context that there is the highest similarity. This process will be iterated until no reassignments occur. The algorithm tends to get stuck in suboptimal solutions [HKD03].

Partitioning Graphs with Coordinate Information

Partitioning techniques for graphs with coordinate information use only the geometric information to partition a graph, the connectivity of the mesh elements will not be regarded [SKK00]. They require the number of clusters to be provided.

> *Typically, geometric partitioners are extremely fast. However, they tend to compute partitionings of lower quality than schemes that take the connectivity of the mesh elements into account. For this reason, multiple trials are usually performed with the best partitioning of these being selected. [SKK00].*

Another way to partition a graph is to find a minimal set of nodes N_{sep} such that the whole set of nodes can be divided into the parts $N_1 \cup N_{sep} \cup N_2$ and such that there is no edge between nodes from N_1 and N_2. N_{sep} is a so-called vertex separator.

One of the basic and most simple algorithms is the Coordinate Nested Dissection (CND) (also referred to as Recursive Coordinate Bisection) approach. A centre of mass of the nodes will be computed and projected onto the coordinate axis that corresponds to the longest dimension of the graph or to a principal inertial axis of the mass distribution. By this procedure, the graph can iteratively be divided into two parts.

[SKK00] highlight that these algorithms are extremely fast, require little memory, and are easy to parallelise, but the partitions tend to have a low quality and contain sometimes disconnected subdomains. Several publications address modifications of the original algorithm and try to deal with the mentioned disadvantages, for example [HR95, NORL86].

Better results show the so-called space-filling curve techniques. There, the single nodes will be assigned to a space-filling curve that fills up space in a locality-preserving way. Therefore, nodes will be ordered and when the curve is split into k parts this results in k subgraphs. Examples for such curves are for example Peano-Hilbert curves. [SW03, ORF96]

Miller et al. propose in [MTTV93] an upper bound for a vertex separator in dimension d. For the set of (α, k) overlap graphs, which includes any planar graph, a vertex separator within a runtime of at most $O\left(\alpha \cdot k^{k^{-1}} \cdot n^{(d-1)d^{-1}}\right)$ can be found. In the algorithm, all points will be projected on a sphere in $(d+1)$ dimensions and a centre-point will be calculated. Then, the points will be mapped and a d-dimensional plane will be created. All points on the intersection points from the sphere with the plane represent the vertex separator.

Multi Agent Systems and Biologically Inspired Approaches

Langham and Grant introduce in [LG99] a graph partitioning algorithm inspired by foraging behaviours of ants. The nodes of the graph are represented by pieces of food and the nest of one of k colonies (each colony represents a partition of the graph). They use a classical bisection heuristic to place the nests in the centre of an equal number of nodes. Ants 'carry' the pieces of foods (the nodes) to an appropriate nest. In the first tests this approach was very competitive on small graphs (up to 286 nodes) and could outperform classic algorithms (for example Recursive Spectral Bisection (RSB), RSB with Kernighan Lin and Multilevel Kernighan Lin). One advantage of this approach is the direct generation of all k partitions without taking the detour over repeated bisection which could be worse for several graphs [ST97].

Comellas and Sapena present in [CS06] a probabilistic algorithm that makes use of ants moving on a graph. Each node of the graph is initially assigned to a random partition. Ants located on arbitrary nodes can now change partition assignments from nodes that produce a high number of edge cuts and change the assigned partition from k_i to k_j with a probability p_{change} to the best or a random partition. Then they search for a node assigned to partition k_j and change this partition to k_i. Due to the probabilistic nature of the algorithm it can escape from local minima. With this approach they were able to be superior to several classic algorithms for special types of graphs.

Another approach by Kuntz and Snyers [KS94] is directly influenced by brood sorting behaviour of ants. They introduced a probability to pick up a vertex from position (x, y), based on function $f(x, y)$ which is an estimate of the vertex density in the neighbourhood.

$$p_{pick}(x, y) = \left(\frac{c_1}{c_1 + f(x, y)} \right)^2$$

In a similar way, a function for dropping a vertex can be defined

$$p_{drop}(x, y) = \left(\frac{f(x, y)}{c_2 + f(x, y)} \right)^2$$

In these formulas, c_1 and c_2 are constants.

Clustering

Another area similar to the graph partitioning problem is the clustering task. In this classic problem field, data will be organised in groups. Clustering algorithms work according to the following two objectives:

Uniqueness: Each data value should only be assigned to a unique cluster
Autonomy: Clusters or categories have to be found autonomously by the algorithm based on the structure of the data

A distinction is drawn between hierarchical and partitioning clustering. The first one finds an indefinite number of clusters by the construction of a tree-like structure with the single data values as leafs and the whole data set as root. Hence, this is far away from the focus of this thesis. The latter splits the whole data set into a given number of clusters. In each clustering process, the similarity between the data values in each cluster has to be maximised. This similarity can be computed by different criteria, one is the minimisation of the quadratic error. Independent from this criterion, there is a huge number of possibilities that have to be compared to find an optimal solution. The number of possibilities to arrange n data values onto m groups is

$$G(n, m) = \frac{1}{m!} \sum_{i=1}^{m} (-1)^{m-1} \binom{m}{i} (i)^n$$

In [HKD03] Handl, Knowles, and Dorigo give an extensive comparison of Ant-Based Clustering (ABC) and the k-means, Average Link and 1D-SOM algorithms with regard to six different evaluation functions. The basic idea of these biologically inspired algorithms[3] is to use ants to group elements in an environment by picking them up and dropping them with different probabilities depending on the size of already grouped clusters. The environment serves as a stigmergic variable. In several algorithms, agents move over a torodial square grid on which data objects are located. Agents pick up these data objects with high probability if it is not surrounded by other, similar, data objects. This is analogue to the idea used for graph partitioning.

---

[3] The clustering of corpses and larvae-sorting activities in real ant colonies [BDT99] are used as an inspiration for these approaches.

ABC shows for nearly all evaluation functions and for both real and artificial test data sets competitive results, mostly it calculates best or second best results. Additionally, the ABC-approach does not need to know the number of clusters in advance since it is able to find them for nearly all types of graphs with a high probability.

3.3.4 Network Flows

If we consider the second objective of our *OPP*, there are obvious similarities with problems from the network flow area. The Multiterminal (or Multiway) Cut Problem (MtCP) (c.f. [DJP+94]) asks for a minimum weight set of edges that separates each terminal from all the others. The MtCP is an extension of the well-known 'min-cut/max-flow' problem. It can be seen as generalisation since the minimal cut deals with two terminals and MtCP with an arbitrary number of terminals.

Definition 10 (The Multiterminal Cut Problem). *Given a graph $G = (V, E)$, a set $T = \{T_1, T_2, ..., T_k\}$ of k terminals, and a positive weight function w that assigns a positive weight to each edge $e \in E$. The question is to find a minimum weight set of edges $E' \subseteq E$ such that the removal of E' disconnects each terminal from all the others.*

Theorem 10. *The Multiterminal Cut Problem is NP-complete for $k \geq 3$.*

Dahlhaus et al. presented in [DJP+94] a nice proof that shows that MtCP is NP-complete for arbitrary graphs. They show it only for $k = 3$, but they point out and it is easy to see that it holds for arbitrary $k \geq 3$. It is done by reducing the Simple Max Cut (SMCP) problem to MtCP. The edges in the SMCP graph will for the transformation be replaced by a so-called Gadget Graph (see figure 3.1(a)).
We will show that a solution for the MtCP from a full-choice OPP graph is an optimal solution to the distance objective of the *OPP*.

Definition 11 (Full-choice OPP graph). *The full-choice OPP graph $G_O = (V, E)$ is a structure representing the distances from each node (agent) to each terminal (target) by edge weights. $V = V_a \cup V_t$ contains agent nodes (V_a) and terminal nodes (V_t). For an OPP instance $OPP = (\mathcal{A}, \mathcal{T}, \rho)$, each vertex v in V_a represents one agent $a \in \mathcal{A}$ ($v := a$, with $V_a = \{v_1, v_2, ..., v_{|A|}\}$, $A = \{a_1, a_2, ..., a_{|A|}\}$). The targets are represented as nodes in V_t, $v_i := T_i$, with $V_t = \{v_1, v_2, ..., v_{|T|}\}$, $\mathcal{T} = \{T_1, T_2, ..., T_{|T|}\}$. E is defined as $\{(v_a, v_t)|v_a \in V_a, v_t \in V_t\}$ and assigns weights to the edges with regard to the distance from the agents to each target, hence $w(v_{a_i}, v_{T_j}) = \delta(a_i, T_j)^{-1}$. The nodes are no longer located in a Euclidean space, the distances have been transformed to edge weights. An example of such a graph is presented in 3.1(b).*

Theorem 11. *A solution for the MtCP from a full-choice OPP graph is an optimal solution for the distance objective of the OPP.*

Proof. A solution for the MtCP is a set of edges E' with minimal weight, that, if removed, disconnects each terminal from all the others. Without loss of generality we consider the terminal $t = T_1$. It belongs to a set of vertices $V' = \{v | (v, t) \in (E \quad E')\}$. We assume that V' is not the distance optimal set of nodes (agents) assigned to t. Then there must be at least one node n that has an edge e' to a non-nearest terminal node. If we denote the edge of this node n to the nearest terminal by e'' which has been removed, we can compare the weight sets of $E_1 = E'$ and $E_2 = (E' - \{e''\}) \cup \{e'\}$. Since $w(e'') > w(e')$, the summed weight of all edges in E_1 is w_1 and of all edges in E_2 is $w_2 = w_1 - w(e'') + w(e') < w_1$. This is a contradiction to our assumption that E' is a minimal cut. □

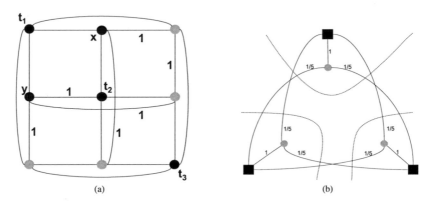

Fig. 3.1. The 'Gadget Graph' (left) is used to proof that MtCP is NP-comlete for $k \geq 3$. The figure on the right side shows an example for a MtCP which is a solution to the *OPP* distance restriction

3.4 Formal Model

We describe the *Online Partitioning Problem* with the formal model proposed by Guttmann et al. in [GZ04]. The problem, or 'Collaborative Scenario', S is a quadruple

$$S = \langle E, T_E, A, P_A \rangle$$

denoted by $ETAP$, that can be extended to the form

$$S' = \langle \langle O, L \rangle, \langle EC_T, MS_T \rangle, \langle IR, M, RA \rangle, P_A \rangle$$

with

- E: environmental state space
  - O: set of objects
  - L: set of locations
- T_E: task specification
  - EC_T: evaluation criteria
  - MS_T: milestones
- A: group of agents
  - IR: internal resources
  - M: models of other agents / groups of agents
  - RA: reasoning apparatus
- P_A: policy of the agents

For our problem the tuple can be filled the following way:

Environmental State Space E

E is a tuple $\langle O, L \rangle$ with

$$O = \{T_1, T_2, ..., T_k\} \cup \{a_1, a_2, ..., a_n\} \Leftrightarrow \mathcal{T} \cup \mathcal{A}$$

$$L_1 = \{\rho(T_i) | i \in \{1, ..., |\mathcal{T}|\}$$
$$L_2 = \{\rho(a_j) | j \in \{1, ..., |\mathcal{A}|\}$$
$$L = L_1 \cup L_2$$

Task T_E

T_E is again a pair $T_E = \langle EC_T, MS_T \rangle$ with

$$EC_T = \{maximise : \prod_{i=1}^{|\mathcal{S}|} |S_i|, minimise : \sum_{i=1}^{|\mathcal{S}|} \sum_{a \in S_i} \delta(a, T_i), minimise : \text{abilities}\}$$

$MS_T = \{ms_0\}$, with $ms_0 =$ "each agent has chosen one target (each agent is stated)"

Group of Agents $\mathcal{A}$

All agents have to optimise each evaluation criteria $ec_k \in EC_T$ to reach the unique milestone $ms_0 \in MS_T$. Unfortunately, in Guttmanns model the different evaluation criteria cannot have different priorities. Therefore, the problem cannot be modelled in detail. Each agent $a_i \in \mathcal{A}$ can be described by the triple $a_i = \langle IR, M, RA \rangle$ where the internal resources IR and models M can be modelled once and remain static because we have a set of identical agents with identical abilities. The reasoning apparatus RA is of no importance because the strategy of the agents cannot be changed during simulation.

Policy P_A

The policy P_A depends on the different strategies. Guttmann understands policy as hierarchical dependencies among agents. In most of the cases there is no need for such a coordination mechanism because each agent decides only depending on local information and without any kind of authority. Only in the emergent organisations presented in chapter 6 there is some kind of hierarchy.

We do not need the whole original $ETAP$ model from Guttmann. Instead we can use the simplified form:

$$S_s = \langle \langle O, L \rangle, \langle EC, MS \rangle \rangle$$

where MS is no longer a set but a single objective.

4

Basic Strategies

Everything should be made as simple as possible,
but not simpler

Albert Einstein

In the previous chapters, we introduced the *OPP* and presented an exact algorithm with a very bad runtime. Hence, in the remaining part of this thesis several heuristics will be presented. The starting points will be the very simple and intuitive algorithms presented here. Some of them turned out to show very good behaviour when the strategy parameters are chosen in an appropriate way.

The strategies are grouped into two main categories, starting with very simple ones that do not use any kind of communication. For these, several worst cases could be discovered. In the second part, the ability to sense and / or communicate with other agents is added. With this additional ability, the results can be enhanced significantly. One approach will turn out to be close to the optimum in randomly distributed settings, even though worst case scenarios where it behaves badly will be identified.

Parts of this chapter appeared in [GBPW05a] and [GBPW05b].

4.1 Non-Communicative Partitioning Strategies

In this section we briefly present heuristics that obviate any kind of communication between the agents. The final state and the quality of the solution depend highly on the initial distribution of the agents. In the affiliated result section these approaches will be compared with the communicative ones.

4.1.1 Random Target Strategy (RTS)

One very simple idea that directly arises when looking for solutions of the *OPP* is the random assignment of agents onto the targets. In the RTS algorithm, each agent decides individually by a simple random operation which of the existing targets it will head for. A similar idea has been used by Comellas and Sapena to find an initial solution for a graph partitioning algorithm [CS06].

In large sets of agents this strategy generates a close to optimum partitioning of the agents onto the existing targets if the random function creates uniformly distributed numbers. But the summed distances to the targets can be significantly higher compared to the optimal solution. In the worst case, every agent can choose the target which is farthest away.

Worst Case Scenario

A worst case scenario can be easily constructed. Suppose we have n agents with the same target T_i as their closest target. By using the RTS algorithm there is a probability $p_{wc} > 0$ that all n agents choose target $T_j, i \neq j$. Hence, the distribution error is 100% and the distance can be arbitrary high, depending on the distance between T_i and T_j. Another example, where the distance error is high, is depicted in figure 4.1.

4.1.2 ID-Dependent Strategy (IDS)

This strategy is similar to the *RTS*. However, the targets will not be chosen by a random number but by the (unique) ID of each agent, according to the simple equation

$$\tau(a_i) = id(a_i) \ \text{MOD} \ m \quad \text{or} \quad \tau(a_i) = i \ \% \ m$$

where m represents the overall number of targets.

This can be reasonable if the agent-IDs are disjoint and complete without any gaps. In this case each target will obtain the same number of agents in the final state (if the number of agents is a multiple of the number of targets). But the same problems as in the random strategy can occur; the value of the summed distances can be very high. We also have to take into account the fact that in a dynamic and / or very large multi agent system the problem of distributing unique numbers onto a changing set of agents is a nontrivial problem and requires more complex agents than for the first strategy. Some ideas for such an ID distribution can be found in [Lyn96].

Worst Case Scenario

In random settings with only a small difference between the number of agents and the number of targets, this strategy has the advantage that the distribution error is always zero but the distance error can still be arbitrarily high. The RTS worst case can be applied to IDS, too.

4.1.3 Next-Target Strategy (NTS)

In this strategy every agent chooses the nearest target compared to its actual position. For this reason it must have an additional ability: the possibility to compare the distances to every possible target. This comparison does not have to be exact, only the relation $\delta(a, T_i) < \delta(a, T_j)$ $\forall 1 \leq i, j \leq |T| = m$ has to be computable by each agent a locally. This can be realised for instance by the stigmergy concept that is described in section 2.1.4.

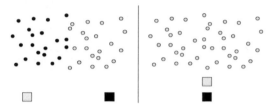

Fig. 4.1. Some exemplary worst case situations for RTS and NTS. On the left side, all agents have chosen the wrong target with the RTS algorithm, hence the distance error is maximal. The right picture shows the worst case situation for NTS, the distribution error is 100% since all agents acting according to this algorithm will choose in each case the light grey target.

Worst Case Scenario

There will be an extremely non-uniform distribution of the agents if the average starting position of the agents is not equidistant to each target. In the worst case, all agents choose the same target. Assume, one target T is closer to all agents than target T'. Due to our algorithm, all agents will choose target T and therefore the distribution error is 100%. This example is presented in figure 4.1 on the right side.

4.2 Communicative Partitioning Strategies

When using the non-communicative strategies the results depend highly on the initial distribution of the agents and the position of the targets in the space. Therefore, we extend the abilities of the agents by some (restricted) communication skills. This section will present some concepts that make use of inter-agent communication in a more or less extensive way. In [Mat98] Matarič discusses some advantages of using communication in multi agent systems to reduce locality by addressing two key problems, the hidden state and the credit assignment problem.

4.2.1 Neighbourhood-Based Strategies (NBS)

In these NBS strategies two factors influence the target selection of the agents. These two factors can be called *group preference* and *individual preference*. The first is based on the number of agents in the neighbourhood[1] heading for each target. The more agents head for one target, the less attractive it is for the currently planning agent. For this calculation, the size of the neighbourhood has much impact on the quality of the decision.

The second factor can be seen as some kind of egoism of the single agents. If the agent is very close to one target, it will prefer this one independent of the preferences of its neighbours. The result is that an agent switches its plan fast if it is far away from any target and the majority of its adjacent agents have the same target. If the agent is very close to its currently chosen target, the agents in its neighbourhood have minor influence on its decision. Initially, all agents choose the nearest target. The concept of *social fields* in multi agent systems is based on similar ideas [BDMA02].

Therefore, the decision for a target depends on the following equation:

$$avoidTarget_i(a) = \frac{1}{k} \cdot |\{b|(b \in \aleph_a) \wedge (\tau(b) = T_i)\}| \cdot \frac{\delta(a, T_i)}{\sum\limits_{j=1}^{m} \delta(a, T_j)}$$

where m is the number of targets and $\{b|(b \in \aleph_a) \wedge (\tau(b) = T_i)\}$ is the set that contains all agents in the neighbourhood of agent a that head for the ith target. k is the size of the neighbourhood. The algorithm chooses the target with the minimal avoidance value.

$$\tau_a = \min_{i=1...m} (avoidTarget_i(a))$$

This calculation will be done by every agent locally until neither the decisions of the agents in the neighbourhood nor the own plan changes. Since this algorithm shows some strange behaviours when applying it synchronously to the set[2], this basic idea will be extended by some additional techniques listed in the following paragraphs.

Neighbourhood-Based Strategy with Delay (NBS_d)

This strategy is an enhancement of the simple NBS. The decision for a target will be renewed only with a small probability (for example $p = 0.01$). One problem with this technique is the determination of this probability and the fact that it is always the same, independent from the number of choices that have been done before.

Neighbourhood-Based Strategy with Annealing Delay (NBS_a)

To modify the probability of the NBS_d algorithm we can use a technique similar to simulated annealing to adjust the target changing probability over time. Starting with a high probability this value decreases every time the plan changes. In detail, the probability starts with $p = 1.00$ and will be halved after each execution. This allows the algorithm to find a stable state fast, but this state might be a local optima. Due to

---

[1] We use the k-neighbourhood $\aleph_a$ introduced in Definition 5

[2] If all agents are acting synchronously, too many of them are changing at the same time. In the next step, all of them will change at the same time to a different target and so on...

the probabilistic behaviour, the algorithm is able to find the global optima and does not necessarily get stuck in a local one.

4.2.2 Border Switch Strategy (BSS)

The strategies we presented before operate without having much special problem-dependent knowledge for the *OPP* problem domain. In case of the BSS algorithm we will take a closer look on the distribution of the single agents. Therefore, we have to extend our assumptions by (non-trivial) additional information observable for every single agent. In this approach, each agent knows the order of the targets, sorted depending on the number of agents addressing this target. This additional knowledge can be motivated by the stigmergy concept introduced by the biologist Grassè in 1959 [Gra59] (see section 2.1.4 in the Definitions chapter). Bonabeau and other researchers [BDT99] from the Santa Fe Institute used this concept successfully to create 3D patterns and other natural phenomena. The targets can be seen as sending individuals and the *stigmergy* (sending strength) decreases when the number of agents that choose this target increases; a similar concept is used in [PGW05] where we examined cooperative agent routing in virtual environments.

The main idea of this algorithm is that only agents located in the middle between two or more targets should adjust their plan with high probability. Whether a single agent is in the middle can be decided by itself from local information obtained by other nearby located agents. Such agents are called *border agents*. The 'middle' between targets is not a position depending on the distances to targets (qualitative) but on the number of agents having chosen these targets (quantitative).

BSS Algorithm

We start with a Proposition stating that there is always a border agent in connected graphs.

Proposition 1. *Assume we have a k-connected neighbourhood graph $G = (V, E)$, a set of minimum stated agents $\mathcal{A} = \{a_1, a_2, ..., a_n\}$, a set of targets $\mathcal{T}$ with $|\mathcal{T}| = m$ and m non-empty sets $Q_1, ..., Q_m \subsetneq A$ with $Q_k = \{a_i | \tau(a_i) = T_k\}$ $\forall 1 \leq k \leq m$. Then there exists a non-empty set of border agents B.*

Proof. Assume the set of border nodes is empty. Then there is no edge $e \in E$ between any of the (by definition non-empty) sets Q_i. If we choose arbitrary $q_1 \in Q_1$ and $q_2 \in Q_2$ then there exist no path $P_{q_1, q_2} \subset E$ that connects q_1 with q_2. This is a contradiction to our premise that G is a kCNG.   □

Our Algorithm 6 works as follows. Each agent decides if it is a border node. If so, it will change its target if necessary. A target flip of agent a_i at time t_j is initiated by recognising that - at the current time - the distribution of all agents is not uniform and the agent a_i had decided for a target stated more frequently. The decision if the

---

Algorithm 6 The Border Switch Strategy algorithm

```
 1:  procedure BSS(Agents A, Targets T)
 2:      p ← 1
 3:      for all (agents a in Population (parallel)) do
 4:          a.choseNearestTarget()
 5:      end for
 6:      repeat
 7:          for all (agents a in Population (parallel)) do
 8:              if (a.isBorderAgent()) then
 9:                  current ← getOwnTargetStigmergy(ℵ_a)
10:                  small ← getSmallestTargetStigmergy(ℵ_a)
11:                  if (current > (small + 1)) ∧ (random() < p) then
12:                      a.changeTargetToSmallest(small)
13:                      p ← p/2
14:                  end if
15:              end if
16:          end for
17:          updateGlobalKnowledge()
18:      until (end condition fulfilled)
19:  end procedure
```

---

agent will change its plan is influenced by a probability value. In lines 3 - 5 the agents initialise their stated target with the nearest one. In lines 9 - 15 the border agents start acting. They compare the assignments to the different targets, regarding only the targets that at least one of their neighbours has been chosen. Thus, they only pay attention to every target that is stated more then one times by agents in $\aleph$. If there is a difference greater than 1 and the state of the border agent is from the larger group, the agent will change its target to the smallest group. We introduced some techniques similar to simulated annealing. The probability to change the targets is decreasing with every new assignment. This is realised by the modification of the variable p in line 13. After the agents' actions, the global knowledge that offers information about the assignments will be updated.

BSS with Global Update

One problem of this approach is that the decision of each border agent which new target it might choose is that the global information might be out-dated. Hence, we add a modification to this strategy that automatically updates the global knowledge of each agent. This is performed everytime a change in the agents' decision occurred. In a real system this could lead to an immense amount of control and information messages, but we added it for comparison reasons. For that reason, line 17 can simply be positioned after line 13.

4.2.3 Exchange Target Strategy (ETS)

Within this strategy the advantages of two non-communicative strategies will be combined with inter-agent information exchange. The RTS has obviously a low distribution error and the NTS offers good solutions regarding the distance aspect.

In its initial state, each agent chooses a random target; therefore, the agents are initially quite uniformly associated with the single targets. But in most initially cases there will be summed distances that are far away from the optimum[3] because the agents' positions have not been set in relation the targets' positions. During the simulation this unbalance is compensated by communication with other agents that are in the neighbourhood. When an agent recognises that there is an adjacent agent that has chosen a different target and a permutation of the chosen targets would decrease the overall distance sum, both will swap their targets. Hence, the distribution quality will be maintained. For the decision whether a permutation will improve the total result, the agents have to be able to compare distances.

Exchange Target Strategy with Probability (ETS_p)

We extend the basic strategy by introducing a probability value p that governs the target interchange between two neighboured agents. In detail, if two agents decide to swap their targets, this will be done with a probability $(1 - p)$. This modification can be seen as a probability delay of agents' actions. Hence, the whole system will not automatically run into local optima where it could not escape from in later steps.

This ETS algorithm cannot perform optimally in every situation; there are several possibilities to get stuck in some kind of local minimum. Figure 4.2(a) visualises

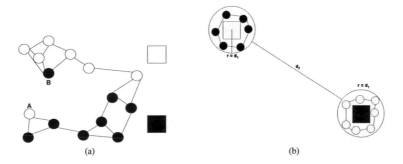

(a)   (b)

Fig. 4.2. Figure (a) shows one situation that origins from a random agent initialisation. Agents A and B will not swap with any of their neighbours. Figure (b) gives an example for a worst case setting for ETS.

this. Because of the missing global knowledge and the missing communication link between agent A and B, both agents will choose the 'wrong' target and therefore increase the overall distance sum.

Exchange Target Strategy with ID Target Choice (ETS_ids)

As we do in the IDS algorithm, we can influence the initial target decision of the agents in the ETS algorithm by unique and consecutive IDs of the agents. Then we

---

[3] compared to the optimal partitioning with highest fitness

can guarantee the uniform distribution of the agents onto the targets, but the worst case remains the same since we cannot influence the overall distance by this policy.

Worst Case Scenarios

For the ETS algorithm, a worst case scenario for the sum of distances compared to an optimal solution can be constructed intuitively. For example, this can occur if there are independent sub graphs that contain only agents having chosen the same target.

Proposition 2. *We can construct an initial graph for any sum of distances, therefore the part of the fitness function representing the distances is unbounded.*

Proof (informal). Assume we have an initial setting as shown in figure 4.2(b). Because of the random target selection, each agent next to the left target (less then d_1 steps away) could have chosen the right target and vice versa. If the distance d_2 between the two not connected sets is noticeably greater than the distance of the single agents to the closest (but not chosen) target ($d_2 \gg d_1$), the agents can have an arbitrary high additional amount of distances. The value of these additional distances is at least $c = \#agents \cdot (d_2 - d_1)$ while the optimal solution is bounded by the value $o = \#agents \cdot d_1$ when the two sets do not communicate at any point of time. By choosing an arbitrary large d_2 we can produce an arbitrary high amount of summed distance. □

Proposition 3. *For n agents and m targets the probability that all agents choose the same target is $p = \frac{1}{m^{n-1}} > 0$.*

Proof. The first agent can choose an arbitrary target and all other $n - 1$ agents can choose this specific target with probability $\frac{1}{m}$. □

Due to this Proposition there is a (very low) probability that the agents distribute themselves in the worst matter. In the remaining part of this thesis we mainly concentrate on the average behaviour of our strategies.

Average Case Scenarios

Due to the uniform random number generator used for the target assignment to agents, there is a very low variance in the distribution of the agents onto the single targets. The values of the initial distribution will not change during runtime of the algorithm. Hence, we only have to predict the distance objective of the *OPP*. Up to now we have three obvious propositions:

- If there is no target swap between any pair of agents during simulation, the ETS algorithm would produce the same results as the RTS algorithm.
- In the best case this strategy can produce optimal results, if the 'right' agents communicate with each other.
- Each appropriate target swap will reduce the overall distance sum and will thus improve the solution quality.

As a direct result from these propositions we can conclude that an increase of communication lines to neighboured agents will reduce the overall distance sum. An optimal solution can be found in a complete graph if enough interaction rounds are provided since every agent can exchange targets with every other one. Hence, all possible combinations can be examined.

4.3 Results

There are several parameters that influence the single strategies. In our simulations, we first compare the strategy characteristics for different numbers of targets. The number of agents has been fixed to a value of 5000, if not mentioned. The other, strategy dependent, parameters will be examined in special simulations. We start herewith and we then compare the individually good parameterised strategies with each other.

4.3.1 Abilities

The introduced strategies cannot only be compared according to their solution quality for the *OPP*. The respective requirements are highly varying. In the following table 4.1 the necessary abilities of the agents are listed. The complexity of the abilities is

Abilities		RTS	IDS	NTS	NBS_x	ETS	BSS
int	`this.currentTarget;`	X			X	X	X
int	`this.uniqueID;`		X		(X)	(X)	(X)
int	`this.consecutiveID;`		X				
double	`getDistanceToTarget();`			X	X	X	X
int	`getTargetOfNeighbour();`				X	X	X
int	`setTargetOfNeighbour();`					X	
int	`getMostVisitedTarget();`						X

Table 4.1. Agent functionality description

ascending in each block. Storing the current target inside each agent needs only a very small memory, but interacting with other agents, identified by a unique ID or an intelligent protocol, or even having global knowledge, demands much more technical and / or computational power. The enhanced strategies need an ID for the communication protocol. This can be generated on the fly by choosing a random number from an interval $\mathcal{I}$ such that it is unique with a high probability for large $\mathcal{I}$. Examples for such interaction protocols can be found for example in [Fer99]. The requirements for consecutive IDs are very high, there have to be some central instance or global communication that distributes these IDs to all agents.

These requirements have to be considered when comparing the outcome of the algorithms in the next sections.

4.3.2 Individual Strategies Tuning

The strategies which do not use communication do not have parameters to be optimised. Hence, in the next sections we concentrate on NBS and ETS parameters.

Tuning NBS Algorithm

Adopting the NBS algorithm, decisions of single agents depend on the decisions of their neighbours. Hence, it is interesting to see the influence of the number of neighbours and the quality of the strategy. In the same algorithm, one can decide to initialise the agents with the closest or with a random target. We examined the behaviour of our strategy for several combinations of these parameters. The results are shown in figure 4.3. Two obvious results can be directly drawn from the experimental results. First,

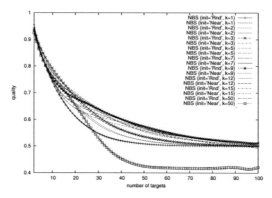

Fig. 4.3. Parameter tuning for the NBS algorithm. We modified the initial assignment of the agents and changed the size of the neighbourhood a decision is based on. The different results are hard to distinguish; the main proposition of this figure is the fact that there is not a big difference in the influence of the different parameters. Only very large or very small parameters show bad behaviour. Some runs will be examined in more detail in figure 4.4.

the way of initialisation has no significant influence on the quality. Similarly, the size of the neighbourhood seems to be from minor relevance. This algorithm demonstrates that large neighbourhood sizes can reduce the overall quality. The best result for most of the number of targets (in the range from 20 to 100 targets) can be reached with $k = 12$. $k = 3$ shows good results for 20 or fewer targets. This strategy is a good example for the possible misinterpretation of additional information obtained by a large neighbourhood. From the experimental runs, a function could be derived that gives for any number of targets x the appropriate size of the neighbourhood $k(x)$ by

$$k(x) = 12 \cdot \left(1 - e^{-\frac{3x}{100}}\right) \ \forall x \geq 1$$

A good compromise for the neighbourhood size is a k value between 9 and 12; this can be chosen for all numbers of targets.

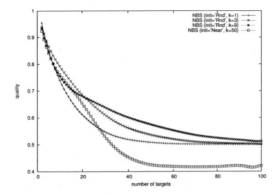

Fig. 4.4. Parameter tuning for the NBS algorithm. In this figure, a selection of all curves from figure 4.3 have been made. The best, the worst and two intermediate parameter settings have been chosen.

Tuning ETS Algorithm

For the ETS algorithm, the initialisation with the nearest target makes no sense, since the primary distribution is a requirement for a good overall quality. But we have an additional parameter, the probability to change the target. Whether this enhances (i.e. downsizes) the overall distance is the focus of this examinations. The influence of very small and very large probabilities is examined. The results in figure 4.5 can be interpreted as follows. If we have a small sized neighbourhood, the algorithm very

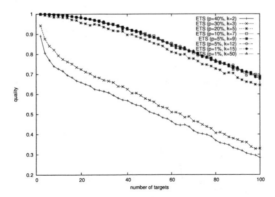

Fig. 4.5. Parameter tuning for the ETS algorithm. We modified the probabilities and the size of the neighbourhood. For each k, the probability with the best results is presented. This is only a small selection of all performed simulations.

fast can become stuck in a local optimum and is not able to leave it while high probabilities have the opposite effect. Then the algorithm will not adjust the necessary target decisions fast enough.

For this strategy, the optimal parameters have to be defined in a different way compared to the NBS algorithm. There, for varying numbers of targets the size of the neighbourhood had to be adjusted. For ETS, there is an optimum value of k for any number of targets, but the probability p has to be adjusted. There is a connection between the size of the neighbourhood and an optimal value for the probability. Hence, for a given neighbourhood size (maybe limited by physical, runtime, energy or economic reasons), the best probability value is needed. In general, the larger the neighbourhood, the smaller an adequate value for the probability can be chosen. Or, the other way round, a lack of information caused by a small neighbourhood can in some areas be compensated by a higher probabilistic manner of the algorithm.

The optimal probability for an arbitrary neighbourhood size could be derived from an extensive number of experiments. The equation

$$\rho(k) = e^{\frac{-2 \cdot x + 29.7}{7}} + 1 \quad \forall k \geq 1$$

represents these results quite good. For any value of $k \geq 1$, it calculates the optimal ETS probability parameter. Both, the experimental data and the approximation function ρ are visualised in figure 4.6.

For the choice of the neighbourhood size, the most reasonable result concerning the *OPP* solution quality is to choose k as large as possible. But this increases the number of inter-agent messages and the maximum in-degree of the agents. In figure 4.5 it is obvious that the solution quality is not rising significantly for neighbourhood sizes larger 7. Thus, this seems to be the best compromise between communication costs and solution quality. Hence, in the later result discussions the ETS parameters will be

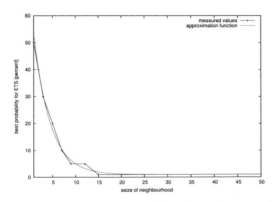

Fig. 4.6. Choice of ETS probability based on neighbourhood sizes. For any k, an appropriate probability value can be computed with an approximation function.

selected according to these results. If no other information is provided, we set k to 7 and p to 10%.

4.3.3 Strategy Comparison

If we compare the strategies as presented in figure 4.7, several conclusions can be drawn. First of all, an increasing number of targets diminishes the quality of all strategies. But the characteristics of the strategies are very different. Since the overall fit-

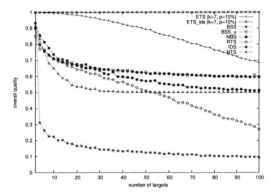

Fig. 4.7. A comparison of the different strategies for a fixed number of agents is shown, dependent on the number of targets. A quality value of 1.0 means that both the partitioning and the distribution objective have been fulfilled perfectly.

ness is a consequence of two single fitness values, the partitioning and the distance fitness can have different individual qualities. For better understanding, these individual fitness curves for all examined number of targets are visualised in the figures 4.8(a) (partitioning) and 4.8(b) (distance).

First of all, the RTS, NTS, and basic BSS algorithm deliver the worst quality results. For RTS it is obvious, the initial high fitness is only based on a more or less uniform distribution of many agents onto few targets, but the distance fitness is far away from the optimal value. Additionally, with a growth of the number of targets, the random distribution becomes worse because the deviation of the random number generator is stronger noticeable. In figures 4.8(a)-4.8(b) both partitioning and distance quality is decreasing with an increasing number of targets. Intuitively, the NTS algorithm has similar problems, but with the other part of the fitness function. The distance is - by definition - always optimal, but the partitioning shows very bad results, especially when dealing with a high number of targets, too.

The reasons for a bad quality are very different for the basic BSS algorithm. Since the decisions of all border nodes are done synchronously and are based on a fixed distribution, the border agents cannot decide for the best new target, they all choose the same target selected by the fewest number of agents. If we consider the enhanced version of this algorithm and update the global knowledge after each decision of a single agent, the results are clearly better. The IDS and BSS_u algorithms show very similar

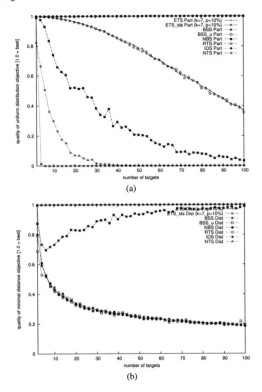

Fig. 4.8. The overall quality presented in figure 4.7 does not describe the quality of the single objectives. Therefore, figure 4.8(a) and 4.8(b) show the quality according to the both objectives, 4.8(a) for the distribution quality and 4.8(b) for the distance minimisation quality. Some curves are overlaid by other ones, in 4.8(a), ETS_ids, BSS_u, and IDS show a constant and optimal quality close to 1.0. Regarding the distance quality, the ETS and NTS algorithms are close to 1.0 and the remaining ones behave similar to RTS.

results, both are able to find an optimal partitioning due to the global knowledge. But at the same time, the distance quality is very bad. The NBS results are somewhere between the IDS and the NTS outcomes, but this is the only strategy whose distance fitness rises if the number of targets increases.

The significantly best algorithms are the ETS strategies. With the parameter setting we have obtained in the previous section 4.3.2, the quality of this approach is better than the quality of all other here presented algorithms for all number of targets. If we give the whole system the ability to distribute unique and consecutive IDs, a constant quality close to 1.0 for any number of targets can be reached.

4.3.4 Standard Deviation

Next, we take a look at the standard deviation of the single strategies. This can give us an idea how varying the quality of the single runs for a setting can be. One result is that the deviation for all algorithms is (acceptably) low, only the NTS algorithm has a higher standard deviation for a small number of targets compared to the remaining strategies. This can be explained by the high dependency of the solution quality of the targets' positions. If there are only a few targets, the probability for them to be uniformly distributed in the whole simulation space is quite low. Hence, the initial distribution has a strong influence on the final result.

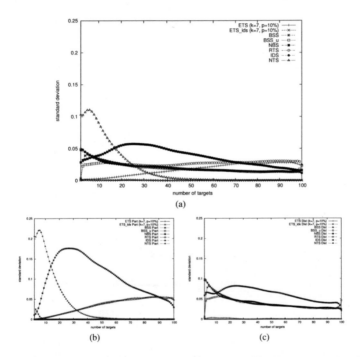

Fig. 4.9. The figures show the standard deviations of the basic strategies depending on the number of targets. Figure 4.9(a) shows the deviation according to the overall fitness, figure 4.9(b) and 4.9(c) extract the deviations for the single objectives.

4.4 Moving Agents

In all approaches presented up to now, the agents decide for a target and assign to it. There is no movement of the agents in space, especially not towards the targets.

But this is an often addressed task in the area of robot coordination. In this section we will examine whether this aspect can be considered with our approaches. For that purpose we address the single strategies and compare if the agents' movements have any (positive or negative) influence.

4.4.1 No Effects through Reconsideration

Most of the algorithms presented before assign to a target based on local information, either according solely to their own position with regard to the targets or by an additional comprehension of the direct and nearby neighbours. When the agents move towards the currently assigned targets, the neighbourhood might change. For NTS, RTS, and IDS, it makes no sense to reconsider the initial decision. The nearest target will remain the nearest one and the decision for a new random target might be better or worse, but the agent is not able to distinguish between such situations; and the IDS algorithm will always calculate the identical output.

4.4.2 Effects through Reconsideration

The situation is different for communicative algorithms. Since the neighbourhood might change and this neighbourhood influences the decision of each agent, reconsideration of the initial decision might be useful. We can distinguish between negative and positive effects of such reconsideration.

Negative Effects (NBS)

The NBS algorithm assigns the agents preferably to the nearest targets. Simultaneously, it divides huge subgroups of agents onto different targets to fulfil the uniform distribution objective. At this point, problems arise when iteratively executing this algorithm on changing neighbourhoods. If we assume that the initial assignment of the agents is not too bad, there are groups of these agents that all move to the same target. When the simulation advances, there is a point in time when all agents only have neighbours assigned to the same target. Since there is no global knowledge and the agents act in an oblivious way, they try to fulfil the uniform distribution objective again. Hence, some agents will be reassigned to other targets. This effect occurs nearby every target. Thus, the distance minimisation objective will not be fulfilled. Some of these problems have been considered and partly solved in [GBPW05b].

Positive Effects (ETS)

There is one strategy which can utilise the effects of moving agents. Since the solution quality of the ETS algorithm is a strictly monotonic function, every executed target exchange among agents has positive effects on the overall solution. The application of the algorithm to moving agents can be described as follows:

Initially, each agent decides for an appropriate target based on a random initialisation and on a local optimisation process in its neighbourhood. After this initialisation, each agent is moving to the currently stated target. If it communicates with another agent on its way and a target exchange would enhance the overall result, both agents swap their target decisions and continue the movement to the new target. Hence, the iterated communication has similar consequences as an increase of the neighbourhood has. We did some simulations and analysed the behaviour in comparison to a solely initial decision.

As a first analysis we will examine what kind of influence the frequency of communication has to the fitness of the final assignment. Therefore, we observe the average fitness for a setting with 1000 agents, 5 targets, and varying communication frequencies. The results are shown in figure 4.10. In the second round of simulations, we

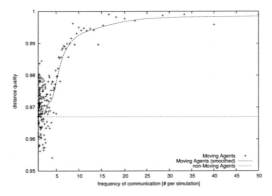

Fig. 4.10. The influence of communication frequency between moving agents. The more often the agents communicate, the better is the overall result, but the influence is decreasing.

analyse the quality of the ETS algorithm for different numbers of targets regarding moving agents. Within this simulation setting, the agents are allowed to communicate 20 times while moving. Since the remaining parameters are the same as in the simulations done for static agents, we can compare the quality of moving agent sets with the quality of static agent sets. In both examinations, the results are very positive even though not surprising. Since the static ETS algorithm produces results close to the maximum - especially when regarding the overall distance minimisation - there are little possibilities for the moving agents to achieve better results. Only when considering smaller numbers of targets (compare figures 4.11 and 4.8(b)), the distance in the static approach is not always optimal. This small lack to the optimum can be overcome for several settings when the agents are moving and hence communicating with a larger neighbourhood. The frequency of communication among agents during movement is from minor importance, as figure 4.10 shows. Communication approxi-

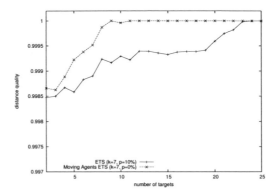

Fig. 4.11. The influence of different numbers of targets among moving agents.

mately 20 times during movement to a target is sufficient to reach an optimal distance value.

4.5 Conclusion

In this chapter, several heuristics for the *OPP* have been presented. Some very simple ones that do not use inter-agent communication have been compared with algorithms for communicating multi agent systems. The non-communicative strategies show sometimes very bad behaviour, especially for problems with a high number of targets. For two approaches, optimal parameters or functions to calculate optimal parameters have been developed. One of these approaches, the *Exchange Target Strategy*, performs best for an arbitrary number of targets and outperforms all other approaches, while it needs only a very small neighbourhood. Thus, the communication overhead is quite small compared to the other ideas. If we initialise the agents based on unique and consecutive IDs, a nearly optimal behaviour of this strategy has been observed. Nevertheless, such an ID assignment implies a central instance or some kind of global knowledge and communication. But even with a random initialisation the performance of this heuristic is very good. For all strategies, some worst cases have been presented that show that these approaches cannot cope with an arbitrary setting.

As an extension, we considered the application of these algorithms to a different setting where the agents move in direction of the target. The ETS algorithm could gain from the additional information given by more inter-agent interactions and improve the quality of the solution.

5

Learning Emergent Agent Behaviours

> *Any sufficiently advanced technology*
> *is indistinguishable from magic*
>
> *Arthur C. Clarke*

Cellular automata (CA) are a very powerful concept and well researched area in computer science and are able to cope with a variety of problems. We will use learning approaches from this field to solve *OPP* instances. These approaches are derived from classification problems in CAs, the so-called *Classification Tasks*. This is a well known and exhaustively studied problem in the field of Cellular Automata [PL03]. In this field of study, a (one-dimensional) binary CA should classify an initial assignment. A more detailed description will be given in the following section 5.1.

In our approach we apply several ideas from the field of the majority classification task for one-dimensional CAs to solve the *Online Partitioning Problem*. One key restriction of the *OPP* is the absence of any kind of central instance. All decisions of the agents are based only on local information, received by communication with direct neighbours. Thus, CAs seem to be a good concept for *OPP* because it respects this restriction as all agents interact with each other exclusively. Therefore, we incorporate ideas of Mitchell et al. [MHC93] and transfer them from one-dimensional CAs to two-dimensional ones. The activated cells in the cellular automata represent the agents. The value of an activated cell describes the currently chosen target according to the *OPP*.

We start with the presentation of a function that handles the mapping of agents, located in a Euclidean space, to a CA assignment. This mapping should take the neighbourhood properties of the agent set into account. While there exist several ideas to embed graphs in data structures or grids (for example [Bor86]), these operate only with the whole graph as input. Here, we present a mapping function in a deterministic and a random version, based only on local information. Moreover, we limit the agents' abilities by introducing inexactness and limited view radii. Hence, our approach is suitable for real world applications with uncertain sensor inputs.

This chapter is organised as follows. Section 5.1 provides an overview of the research in the Majority / Density Classification Task for one-dimensional cellular automata. In section 5.2 the mapping function is presented and motivated. Its quality is shown in experimental runs. In the next section 5.3 we describe our approach denoted by *CAS* as well as presenting an algorithm. The simulation settings and the received results are presented. We enhance the basic idea in section 5.4 by introducing some new evolutionary operators and compare the new results with the basic ones.

The mapping function has been published in [Goe06b] and the rule learning approach appeared in the CEC proceedings [GWP05].

5.1 Related Work

The *Majority Classification (MC)* or *Density Classification (DC)* task is a well known and exhaustively studied problem in the field of Cellular Automata (CA) [PL03], first published by Gacs, Kurdyumov and Levin [GKL78]. A one-dimensional binary CA with an initially random state for each cell $c_i \in \{0, 1\}$ should result in a uniform state for all cells depending on the initial ratio of 0's to 1's. Consequently, if there were more 1's than 0's in the initial configuration, the final state of each cell should be $'1'$, if there were more 0's, the final state of each cell should be $'0'$. We define the final state to be the CA assignment after a predefined and fixed number of transitions.

There exist many different approaches to generate rules for classifying a high proportion of such random initial cell assignments. The best approaches found rules which automatically classify more than 86% of the configurations correctly. Manually generated rules reach a correct classification value of around 82%; the best one has been obtained by Das, Mitchell and Crutchfield with 82.178%. One of the earliest ones can simply be expressed by two rules.

IF $(cell_i(t) = 0)$ THEN $cell_i(t + 1) = majority\{cell_{i-1}(t), cell_{i-2}(t), cell_{i-3}(t)\}$
IF $(cell_i(t) = 1)$ THEN $cell_i(t + 1) = majority\{cell_{i+1}(t), cell_{i+2}(t), cell_{i+3}(t)\}$

Each cell considers either the left or the right neighbours, depending on the current cell state. The state in the next time step is simply the majority of these neighbours.

For the automatic generation of rules for cellular automata, several techniques have been used, for example genetic algorithms [MHC93], genetic programming [ABK96], or coevolution [JP98, PM01, WMC00]. In [Kir05] an exhaustive study of the *Majority Classification Task* has been made[1].

---

[1] There, Kirchner could reproduce most of the results obtained by Mitchell et al. Under our supervision she developed a software tool that can also be used to group the values in randomly initialised one-dimensional cellular automata.

5.2 Mapping Agents to Cellular Automata

We use approaches from the cellular automata research to solve optimisation problems in the multi agent system research area. For this purpose, we require a transformation from agents located in a Euclidean space into an abstract cell assignment for cellular automata. In this section, a mapping function is presented and evaluated with a reverse function. This function can be calculated by each agent individually based only on local information. Additionally, we examine the performance of the function in inexact and non-deterministic environments.

5.2.1 Mapping Function

Our idea presented in this section is to map a set of agents, that are located somewhere in space, onto a two-dimensional cell-grid with two main states, *active* and *inactive*; an active grid field represents the presence of an agent. This grid is the cell assignment of a cellular automaton. The transformation from an agents' position to the grid will be realised by a mapping function, denoted by ψ. It is important to maintain the neighbourhood relations of agents in space in the activated cells in the cellular automata. Hence, agents that are located nearby in space should be direct neighbours in the grid. A major problem for such a function ψ is the decision if an agent is located nearby, because this decision has to be calculated based exclusively on local information.

ψ has to decide for each agent $a \in \mathcal{A}$ with a fixed but only locally known position p in space, how the adjacent cells $\mathcal{N} = \{n_0, n_1, ..., n_8\}$ (including the centre cell n_0 itself representing the current agent) on the CA should look like to represent the original situation as best as possible.

To test the quality of such a function ψ, we need the ideal mapping as a reference value. Therefore, we decided to start with a random cellular automaton $\mathcal{CA}$ and a

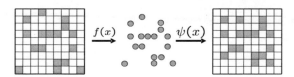

Fig. 5.1. An idea how to measure the mapping-quality

mapping function f from activated cells of a CA to coordinates of agents in space. This agent set is the input for our mapping function ψ. The CA description $\mathcal{CA}_{mapped}$, which is derived from ψ, can then be compared with $\mathcal{CA}$.

The function $f : \mathcal{CA} \to \mathbb{R}\mathrm{x}\mathbb{R}$ is defined by

$$f(c_{i,j}) = \begin{cases} \text{agent on position } (i,j) \text{ if } c_{i,j} \text{ is activated in } \mathcal{CA} \\ \text{no agent on position } (i,j) \qquad\qquad\qquad else \end{cases}$$

This is visualised in figure 5.1.

The idea of our mapping function ψ is to find for each agent a an appropriate representation for the terms 'near' and 'far' and to decide for each nearest agent in one of the eight sectors around a if it is 'near' or 'far'. The first problem is solved by the consideration of the nearest and the farthest among the eight closest agents in the sectors. These two extreme values constitute our normative system. The second part is realised by using a special gradient for ψ. The concept to divide the space into sectors originating at each agent is similar to the construction of efficient graphs for mobile ad hoc networks ($MANETS$). One example of such successful graphs is a special kind of Yao graphs, the sparse Yao graph (SparsY, see [WL02] and [Epp96]). For these graphs, each node divides the space into k cones and draws an edge to the nearest node in each sector. Therefore, the in- and out-degree of each node is limited to at most k.

Deterministic Mapping

We start with the mapping of the (highly environment distance measurement dependent) distances from the agent a to its neighbours in each sector into the standardised interval $[0; 1]$. After the mapping, the distance value of the closest agent is 0 and the distance value of the agent that is farthest away is 1. These distance values are now independent from the original values, so we do not longer have to take into consideration if we are dealing with metric systems like kilometres or nanometres etc. The conversion is established by the equation

$$\text{r-dist}(a, a_k) = \frac{\delta(a, a_k) - \min_{i=1,\ldots,8}\{\delta(a, a_i)\}}{\max_{i=1,\ldots,8}\{\delta(a, a_i)\} - \min_{i=1,\ldots,8}\{\delta(a, a_i)\}} \in [0; 1] \tag{5.1}$$

separately for each agent depending on the nearest agent in each sector[2]. In this equation, the agent a_i is the nearest agent in the sector i, this sector can be described as the cone with an angle of $\left(\frac{2 \cdot \pi}{8}\right)$ which has sides from $\left(\frac{i-1}{8} \cdot 2\pi\right)$ to $\left(\frac{i}{8} \cdot 2\pi\right)$.

Next, we define a strictly monotonic decreasing function ψ.

$$\psi(x) = 2^{-x^2 \cdot \left(\frac{1}{8} \cdot \sum_{i=1}^{8} \text{r-dist}(a, a_i)\right)^{-2}}$$

This is our mapping function that determines for an arbitrary relative distance (r-dist) value $x \in [0; 1]$ if this is a far or a near distance. Characteristics of this function can be found in B.1.

The decision, if an agent is a neighbour or not, depends on the value of this function, and we can decide for each sector if the agent is close enough to appear in the

---

[2] If the denominator of the fraction is zero, we force r-dist to return 0.5 because all agents have exactly the same distance.

neighbourhood of agent a or not. Therefore, the value for each cell $n_0, ..., n_8$ in the neighbourhood of an arbitrary agent a is defined by

$$n_k = \begin{cases} active & \text{for } k = 0; \\ active & \text{for } \psi(\text{r-dist}(a, a_k)) \geq \frac{1}{2}; \\ inactive & \text{else} \end{cases}$$

Random Mapping

The random mapping is a slight modification of the deterministic mapping presented before. We allocate the adjacent cells of agent a with a probability that depends on the function ψ.

$$n_k = \begin{cases} active & \text{for } k = 0; \\ active & \text{with probability } p = \psi(\text{r-dist}(a, a_k)); \\ inactive & \text{else} \end{cases}$$

Measuring the Error Rate

We test our mapping function $\psi(x)$ by comparing the mapping result onto a cellular automaton with the optimal solution. For that purpose, we start with a cellular automaton cell assignment C_1 with a varying density d of activated cells ($d \in \{0\%; ...; 100\%\}$). The activated CA-cells will be converted into a set of agents with position in space based on the cell position in the CA. For cell $c_{i,j}$ the position in space will be (i, j). Then we apply our algorithm to this set of agents and convert them back into a cellular automaton cell assignment C_2. The quality of the final as-

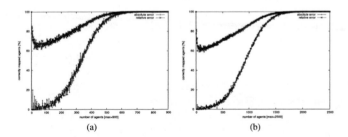

(a)   (b)

Fig. 5.2. The development of the absolute and the relative error metrics for different numbers of agents. The curves are nearly identical, as the agents make only local decisions.

signment C_2 is measured by two error metrics. The absolute error metric represents the ratio of corresponding activated cells that have an identical neighbourhood in C_1

and C_2 to the overall number of activated cells. The relative error regards similarities in the neighbourhood of correspondent activated cells. In figure 5.2 we compare the mapping quality of our algorithm for different CA-sizes. The absolute error err_a compares the Moore neighbourhood of all occupied fields and sets the number of identical assignments in relation to the number of activated fields. This is done by

$$err_a = \frac{\displaystyle\sum_{c \in C_2 \ with \ c \neq empty} \begin{cases} 1 \ for \ n(c \in C_1) = n(c \in C_2) \\ 0 \ else \end{cases}}{|\{c | c \in C_2 \ and \ c \neq \ empty\}|}$$

The relative error err_r measures not only totally identical Moore neighbourhoods but also partly identical Moore neighbourhoods by

$$err_r = \frac{\displaystyle\sum_{c \in C_2 \ with \ c \neq empty} \Delta(n(c \in C_1), n(c \in C_2))}{9 \cdot |\{c | c \in C_2 \ and \ c \neq \ empty\}|}$$

In this equations we use $n(c)$ as a function representing the Moore neighbourhood of cell c (and c itself) and the difference function $\Delta(n, n')$ calculates the number of differences between the neighbourhoods of n and n' with $\Delta(n, n') \in \{0, ..., 9\}$.

These error metrics have been chosen to calculate how close the mapping is to the original situation (err_a) and how the neighbourhood relations can be maintained (err_r).

The results in figure 5.2 show that the development of the error function is similar for different CA dimensions. Hence, we concentrate on agent densities and abstain from the absolute CA size in the coming examinations.

5.2.2 Introducing Inexactness and Limitations

In the previous section we assumed that the distances could be measured with no kind of noise and 100% precision. Now we will look at the behaviour of our algorithm if we add noise to the distance measurements. On account to this, we modify the correct distance value by p percent ($dist_m = dist + m$ with $m \in [-p \cdot dist; p \cdot dist]$). The other parts of the algorithm will not be changed. Figure 5.3(a) visualises that the characteristics of the random and the absolute mapping show a high similarity, but for cell densities greater than 50% the quality of the random mapping is significantly better. The curves show a difference of more than 5%.

Following the success of our mapping function in an ideal environment and satisfying results when adding inexactness, we analyse our approach with more realistic agents, i.e. agents with limited sensor information with respect of the view radius. To be more precise, we observe the behaviour of our ψ while applying different view radii. For this reason, we modify our original equation (5.1) as follows:

$$\text{r-dist}_{vr(r)}(a, a_k) = \frac{\delta_{vr(r)}(a, a_k) - \min_{i=1,...,8}\{\delta_{vr(r)}(a, a_i)\}}{\max_{i=1,...,8}\{\delta_{vr(r)}(a, a_i)\} - \min_{i=1,...,8}\{\delta_{vr(r)}(a, a_i)\}} \in [0; 1]$$

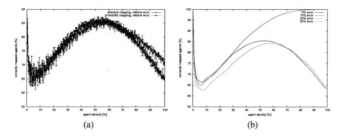

Fig. 5.3. Comparison of the absolute and the probability mapping function when the agents cannot measure distances exactly. In figure 5.3(a), we use an error of at most $p = 10\%$. In 5.3(b), we compare the development for different error sizes.

The difference between the r-dist$_{vr(r)}$- and the original r-dist-function is that we now have a differentiation of two cases

$$
\delta_{vr(r)}(a, a_k) = \begin{cases} \delta(a, a_k) & for\ \ r \gtrsim \delta(a, a_k); \\ \infty & else \end{cases}
$$

In the figures 5.4(a)-5.4(c) we compare the quality of our mapping function for agents using limited view radii r. The influence of the real value r depends obviously onto the real value distances of the agents in the space. For the experiments presented here we chose the value of radius r from a reasonable interval[3]. The experiments show that the choice for the view radius value is from minor importance. We distinguish three different cases:

No Sensing: If the view radius is smaller than the minimum distance of any two agents, the mapping function can obviously map the agent set only into a cellular automaton in which no cell has any neighbour. Therefore, the quality of the mapping function is decreasing when the density of the agents and therewith the average number of neighbours in the original CA is increasing. This can be seen in all settings in figure 5.4 considering the quality development for the view radius $VR = 0$.

Over Sensing: If the view radius is greater than the maximum distance between any two agents, there is no change in the knowledge of the agents because with an increase of the radius no additional information can be gained.

Variable Sensing: If the view radius is inside the reasonable interval, then the quality of the mapping function changes. We can distinguish between two cases. For sparse agent densities, a decrease of r increases the mapping quality. In such settings, agents that are far away can be misleadingly considered as near neighbours

---

[3] A *reasonable* interval $[I_1; I_2]$ is obviously an interval with the minimum and maximum distance of agents in the simulation space as interval borders. Hence, I_1 should be equal to or greater than the minimum distance between any two agents a and a' and I_2 should be smaller than or equal to the maximum distance between any two agents a and a'.

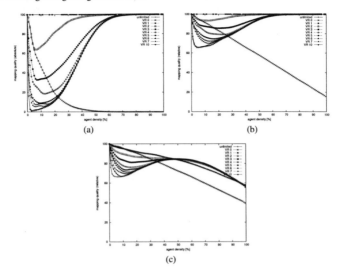

(a)   (b)

(c)

Fig. 5.4. Comparison of different view radii of the agents. Again, we compare the results of our mapping function for different agent densities. The figures 5.4(a) and 5.4(b) show the behaviour of our mapping function without errors in the distance measurement. The absolute and the relative errors are shown. Figure 5.4(c) is the result of additionally increasing the distance inexactness of the agents to 10%.

because there are not enough agents to calculate a good normative distance value base. An example for such a bad case is shown in figure 5.5. For an agent density

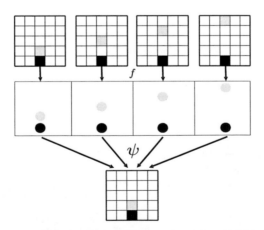

Fig. 5.5. A situation where our mapping function fails. Due to the missing global information, the mapping function can not distinguish between the four originally different cell assignments. For our equation, the grey agent is the nearest one in all settings and therefore always a neighbour for the black agent.

of approximately 50% there is no longer a significant difference for the different radii VR.

There is one special case if the view radius is equal to the minimal distance between any two agents. This radius is identical to the function we use to map the agents from the original CA to the agent set in space. In the experiments with no inexactness, $r = 1$ is a direct inversion of this function and therefore the mapping quality is always exactly 100%. In the experiments with inexactness, there are some agents that could not be sensed because they are out of the view radius. Therefore, we receive slightly worse results.

5.2.3 Conclusion

The presented mapping function to transform a group of agents into active and inactive cells of a cellular automaton shows some promising characteristics. Its quality has been determined by a comparison with the ideal cell assignment obtained via a double transformation process. To make the system more realistic and tolerant, we added inexactness and limited agent abilities. In all combinations, our mapping function ψ shows a behaviour that ranges from acceptable to very good, mostly depending on the agent density in the system. The best results can be obtained if the agents' distribution is not too sparse.

5.3 The Set of Rules for the CA and EA Operators

We will now elaborate on our idea to use cellular automata rules as solutions for an *OPP* instance. Section 5.2 gave an example how to convert agents in space into an assignment of cellular automata cells. We will now present a learning algorithm that finds an appropriate set of rules for arbitrary cellular automata to enhance the solution quality for an *OPP* instance. A rough idea of the mapping for the *OPP* to the density classification task (DCT) problem is visualised in figure 5.6.

5.3.1 Definitions

In this chapter we make use of the following terms:

Individual An individual I is defined as a list of rules

$$I = (R_1, ..., R_k)$$

Rule A rule $R = (c, M, D, r)$ is defined as a single transition step for the CA. The input for a rule is the value of the occupied cell $c_{x,y}$, the Moore neighbourhood $M = \{c_{x-1,y-1}, c_{x,y-1}, c_{x+1,y-1}, c_{x-1,y}, c_{x+1,y}, c_{x-1,y+1}, c_{x,y+1}, c_{x+1,y+1}\}$; $|M| = 8$ of this cell with $c_{a,b} \in \{1, ..., |T|, \emptyset\}$, a distance information $D = \delta(c_{x,y}, T)$ with *distance granularity* $|T|$ to this target (see below), and a new value (rule result) r for the cell.

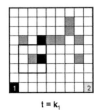 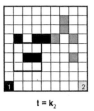

$t = k_1$ $t = k_2$

Fig. 5.6. Mapping the *OPP* to the *Majority Classification Task*: Two points in time (k_1 and k_2 with $k_1 < k_2$) are visualised here, the agents are represented by active cells in the two-dimensional cellular automata, the different grey values indicate the currently chosen target (the targets are here T_1 and T_2, located on the lower edges). The number of agents will not change in any point in time (see footnote 5). To map the *OPP* to the CA, we simply map the agents onto fields of a Cellular Automaton.

Distance Information The centre cell has - in addition to the information about the neighbouring cells - information about the distances to the targets. This information is provided as a vector $V = (v_1, ..., v_{|\mathcal{T}|})$ where $v_i \in \{0, ..., |\mathcal{T}|-1\}$ represents the relative distance to the target T_i. The closest target obtains the relative distance 0, the target which is farthest away obtains the relative distance $|\mathcal{T}| - 1$. The remaining targets will be assigned with the numbers between 0 and $|\mathcal{T}|$ according to their distance. $V = (0, 2, 3, 2)$ implies that target T_0 is the closest target, the distance to T_1 and T_3 is - relatively seen - the same and target T_2 is farthest away.

5.3.2 Evolutionary Operators and Fitness Function

We use an evolutionary algorithm (EA) (for an introduction in evolutionary algorithms, see section 2.1.5) to find useful rule combinations. For the EA we have to specify several operators and a fitness function. Concerning the population structure and the selection scheme of our evolutionary algorithm we use a (μ, λ)-ES (evolution strategy). The size of the parental population is μ. In each generation λ offspring individuals are produced applying the variation operators *mutation* and *crossover*.

Mutation

The mutation operation mut alters the result and / or cell patterns in the neighbourhood of the original rule.

$$mut(R) = R'$$

We use different probabilities for executing the mutation operator. For each rule the rule mutation probability m_r describes the probability to be mutated. The cell mutation probability m_c defines the probability for the cells in the neighbourhood, for the centre cell, an entry in the distance vector and for the resulting values to be mutated. Hence, there is a (very small) probability that more than one mutation operation in a single rule occurs.

Crossover

We implement a single point crossover operation $cross$. If $p \in \{2, ..., k\}$, uniformly and randomly distributed, is the crossover point, two new individuals will be created by the formula:

$$cross(I_1, I_2, p) = \{I_3, I_4\}$$

with

$$I_1 = (R_1, ..., R_k), I_2 = (S_1, ..., S_k)$$
$$I_3 = (R_1, ..., R_{p-1}, S_p, ..., S_k)$$
$$I_4 = (S_1, ..., S_{p-1}, R_p, ..., R_k)$$

Fitness

The selection process of the EA is based on the fitness values of the single individuals. The fitness itself is an evaluation of the resulting distribution of the agents onto the targets and the summed distances to them. In the cellular automata representation the agents are represented by activated cells in the CA.

For the fitness function we simply use our equation 3.1 that describes the quality of the *OPP* solution.

$$f = \alpha \cdot \left(\frac{\prod_{i=1}^{m} b_i}{\prod_{i=1}^{m} o_i} \right) + \beta \cdot \left(\frac{\sum_{i=1}^{n} \min_{j=1..m} (\delta(a_i, T_j))}{\sum_{i=1}^{n} \delta(a_i, \tau(a_i))} \right)$$

A high fitness means that the agent distribution deducted from the cellular automata is close to the optimum solution for the *OPP*.

5.3.3 Rule Similarity

The overall number of rules for handling every possible neighbourhood in a cellular automaton is

$$\underbrace{(|T| + 1)^{|M|}}_{diff.\ neighbourhoods} \cdot \underbrace{(|T| + 1)}_{centre\ cell} \cdot \underbrace{(|T| + 1)}_{result} = (|T| + 1)^{|M|+2}$$

with $|T|$ representing the number of targets[4] and $|M| = 8$ representing the size of the neighbourhood (see section 5.3.1). If we add the distance information, this value will even be increased by the term $5^{|T|+1}$, so that we obtain a set $\mathcal{R}$ of all possible rules with the size

$$|\mathcal{R}| = (|T| + 1)^{10} \cdot 5^{|T|+1}$$

---

[4] This number is increased by 1 in the formula because we additionally have the *inactive* cell state.

If we regard the combination of rules in a set of rules that can deal with any neighbourhood, there are

$$(|T| + 1)^{2^9} = (|T| + 1)^{512}$$

of such rule sets.

As it is difficult to decide in reasonable time which of the rules or combinations of rules in $\mathcal{R}$ are important or useful to solve the partitioning problem, we use evolutionary algorithms to select only a small set of necessary rules $\mathcal{R}'$ out of the whole set of all possible rules. However, with this selection of rules we cannot handle every possible cell neighbourhood. Therefore, we design several distance measures to find the most similar rule to a given neighbourhood.

Value Count Similarity (VC)

With this similarity we compare the number of values in the neighbourhood; the position of the single cell values might be different. The VC-similarity of two neighbourhoods

$$N = (n_1, ..., n_8)$$

and

$$N' = (n'_1, ..., n'_8)$$

with $n_i, n'_i \in \{0, ..., |T|\}$ is defined by

$$\sum_{i=1}^{|T|} min(|\{j|n_j = i; 1 \leq j \leq 8\}|, |\{j|n'_j = i; 1 \leq j \leq 8\}|)$$

Equal Bit Count Similarity (EBC)

This similarity-measure value describes the number of identical values in two Moore neighbourhoods. The EBC-similarity of two neighbourhoods $N = (n_1, ..., n_8)$ and $N' = (n'_1, ..., n'_8)$ is defined by

$$EBC = |\{i|n_i = n'_i, 1 \leq i \leq 8\}|$$

Smoothed EBC (sEBC)

The basic EBC strategy is very stringent and counts only the absolute identity of the cell values. We want to smooth this procedure to allow a more robust similarity value. The similarity between two cells with different values depends in this extension additionally on the values of the cells nearby. Therefore, we convolve the neighbourhood with a Gaussian smoothing filter. This operator is commonly used in image processing and its filter matrix $F = (f_i)_{-r \leq i \leq r}$ is defined by

$$f_i = e^{-\frac{i^2}{r^2}}$$

where r is the radius of the filter. For further information about convolution and Gaussian smoothing filters see [GW92]. We used some similar techniques for comparing matrices in [PKWG05].

We apply the filter for each possible cell value individually. Therefore, we generate for each value $v \in \{1, ..., |\mathcal{T}|\}$ and a neighbourhoods $N = (n_1, ..., n_8)$ a vector $G_v = \{g_{v,1}, ..., g_{v,8}\}$ with

$$
g_{v,i} = \begin{cases} 1, \text{ if } n_i = v; \\ 0, \text{ otherwise.} \end{cases}
$$

and convolve each vector G_v separately with the Gaussian filter. The similarity of this neighbourhood N with another neighbourhood N' is then calculated by the formula

$$
sEBC = 1 - \left(\sum_{i=1}^{|\mathcal{T}|} \sum_{j=1}^{8} (g_{i,j} - g'_{i,j})^2 \right)
$$

This is based on the Euclidean distance between two vectors. An idea how this filter

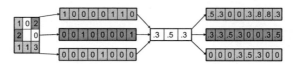

Fig. 5.7. An example how the Gaussian filter works. The one / zero similarities will be smoothed to more fuzzy ones.

works is presented in figure 5.7.

5.3.4 Algorithm & Worst Case Scenario

In the following we present an abstract version of the algorithm we use to create a heuristic that solves the *OPP*. In the second part we show that this algorithm can not handle all possible initial CA configurations in a satisfying way.

The CAS Algorithm

An example how to find successful sets of rules is presented in Algorithm 7. Starting with a random population P, each individual (i.e. a number of rules) in P will be assigned to a random cellular automaton instance. A number of (*caTransition*-many) steps will be made by the CA. Following, the final assignment will be rated with regard to the *OPP*. This reflects the fitness of the individual. When this is done for all individuals, a new population will be generated based on basic evolutionary algorithm operators.

Algorithm 7 The algorithm to find successful sets of rules that solve the *OPP*

```
1:  procedure CAS_OPP_SOLVER
2:      p ← random population
3:      while (not reached STOP_CRITERION) do
4:          for each ruleset R in population P do
5:              generate random CA instance
6:              for (i ≤ #caTransitions) do
7:                  use ruleset R on random CA
8:              end for
9:              rate the partitioning quality according to the OPP
10:             update the individual in population P
11:         end for
12:         new Population newP ← selection(P)
13:         newP ← newP ∪ crossover(newP)
14:         newP ← mutation(newP)
15:         P ← newP
16:     end while
17: end procedure
```

Worst Case Scenario

In this section we will prove that the algorithm we presented in listing 7 cannot handle all possible initial CA configurations to produce results that are close to the optimal partitioning. For the *MCT* on one-dimensional CAs there is a proof in [LB95] for binary and in [CSY99] for n-ary CAs that shows that there exists no rule set solving the *Density Classification Task*. In the next Theorem we will show that there is also no optimal set of rules for arbitrary cell assignments / agent distributions.

Theorem 12. *The distribution error of n agents onto k targets can be arbitrarily close to the highest error value (100%).*

Proof. Assume we have a random initial cell pattern situation as visualised in figure 5.8. In this situation we have n active cells with an empty neighbourhood. In addition, the distance information is identical for all active cells. Therefore, the centre cells can not distinguish between their neighbourhoods and their distance information. In this situation there is only one rule all cells use at the same time. Therefore, all cells will choose the same target and the distribution error is 100%.   □

We will only consider random CA configurations and will evaluate our evolutionary approach on random cell assignments. In our extensive experiments the possible occurrence of the mentioned worst case scenario showed no effect.

In the following section we present some simulation results for the above mentioned strategies. We start with information about the implemented simulation environment, the simulation settings, followed by the description and explanation of the results.

5.3.5 Simulation Environment & Parameters

In order to obtain comparable results, we deliberately use similar parameters to those of the basic strategies in chapter 4. The simulation is split into two phases. In the first

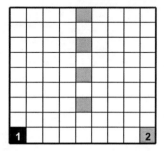

Fig. 5.8. An examplary CA configuration that shows a worst case scenario according to the distribution quality. For all active cells, the targets T_1 and T_2 are equidistant.

phase the evolutionary algorithm is allowed to evolve for at least 1000 generations on small, random cellular automata. In each generation the best individual is chosen to perform on a bigger cellular automaton with $2000(50\%)$ from ≈ 4000 activated cells[5]. This test constitutes the second phase of each simulation. Both automata are toroidal.

5.3.6 Parameter Tuning

In this section different parameter settings for learning rules for Cellular Automata are compared with each other. Therefore, we can investigate which parameters are important or relevant to improve the overall result.

Size of the Ruleset

First, we investigate the influence of different numbers of rules which can be learned as transitions. Figure 5.9 presents the influence of this parameter to the fitness of the whole set for arbitrary cellular automata. It is easy to see that an increase of the number of rules that can be learned will not automatically lead to better results. Quite the contrary behaviour can be observed. Small rulesets which contain around 100 rules show best results. The reason is the size of the solution space. A small number of rules can effectively specialise in reasonable behaviour covering a wide range of neighbourhood assignments. They are able to generalise from and classify an adequate number of different neighbourhoods.

---

[5] we only allow rules that change the state of an active cell to another active state, an active cell can never become inactive and vice versa. This is to ensure that the overall number of activated cells stays the same during the simulation. There exist several papers dealing with density maintenance, for example [BF98, BBP98, BF02], but we chose the straightforward way to maintain spatial distribution characteristics.

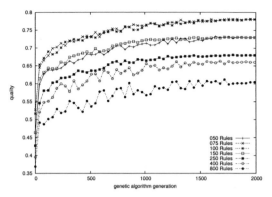

Fig. 5.9. Comparison of the fitness obtained with different numbers of rules. Both, too small and to large rulesets show bad behaviour. In the latter ones, the search space is too large or there are too many possible rule interactions. In too small sets the opposite problem occurs. The single rules can no longer distinguish between a sufficient large number of different neighbourhoods, the generalisation is too imprecise.

Similarity Measurement

The second parameter that has to be chosen is the similarity measurement method. We introduced three different approaches in the preceding sections. Again, the best method for the learning of rules can be derived in several simulations (see figure 5.10). The very basic and imprecise approaches remain static on a low level. The *s*EBC

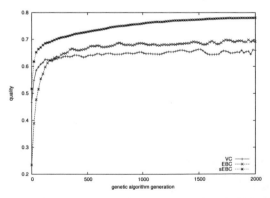

Fig. 5.10. Comparison of the different similarity measurements. Both *value count* and the standard *equal bit count* stagnate on a significantly lower level compared to the enhanced *smoothed equal bit count (sEBC)* filter algorithm.

method which is able to distinguish between similar and non-similar neighbourhoods on a fuzzy level performs significantly better.

5.3.7 Results

In the last sections we present only assignments for two parameters. These assignments are still not necessarily optimal ones since the single parameters are highly connected to each other. But they show a good behaviour in several simulations. The assignment of the remaining parameters is listed in the following table 5.1. Each of them has been modified and adjusted in several simulation runs which are not all listed here. For the evaluation of this approach and the comparison with other approaches,

Parameter	Value
CA size (phase I)	20x20
cells activated (phase I)	200 (50% active cells)
CA size (phase II)	63x63
cells activated (phase II)	2000 ($\approx$50% active cells)
CA transitions	3
#rules in rule set	100
#targets ($|\mathcal{T}|$)	5
similarity	sEBC ($r = 1$)
population size	120
μ	20
λ	100
selection	best (comma)
crossover	one point
rule mutate probability m_r	0.1
cell mutate probability m_c	0.01

Table 5.1. Determined parameters for the rule learning algorithm for cellular automata

this parameters have been chosen. In figure 5.11 we list several basic algorithms for the OPP taken from chapter 4 and compare the overall errors with each other. The CAS algorithm appears to be of medium quality after a sufficient long learning phase. It can outperform the very basic approaches but is far away from the performance of the ETS algorithm. One reason for this might be the huge solution space and the high requirements to rule interactions at consecutive executions.

5.3.8 Conclusion

In this approach, a lot of parameters appeared which all have to be adjusted. Some of these adjustment processes have been presented. The remaining assignments have been found in exhaustive simulation runs. These runs need a lot of time since the evaluation of each single ruleset is very time-consuming[6]. In literature about the majority classification task [MHC93] and in our own simulations we could find out that the

---

[6] 1000 generations need approximately one week of computation on a fast machine.

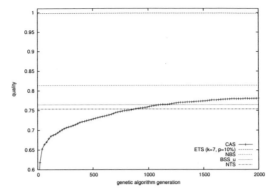

Fig. 5.11. Comparing the CAS algorithm with other heuristics solving the Online Partitioning Problem. To improve readability, we omitted the RTS and the IDS algorithm. They are located with a fitness value around 0.75 between BSS_u and NTS. All the quality values have been taken from the runs presented in chapter 4.

size of the cellular automaton has minor influence on the quality of the final ruleset during the learning phase. The only restriction is not to chose it too small in order to minimise the influence of the random initial distribution. The size chosen here is a compromise between runtime and good average behaviour. This seems to be sufficient since the interactions are designed to be solely local.

5.4 Memetic Individual Modifications

In the former sections we use a evolutionary algorithm to create suitable sets of rules (RS) for cellular automata that could solve the *OPP* successfully. But the identification of successful sets is extremely slow and needs a high amount of computing power. Additionally, we found out that in these RSs often only a small choice of rules will be executed when running on a random cellular automaton or *OPP* instance. Most of the containing rules will be used very seldom. One solution would be to reduce the size of RSs, but then we directly influence the exploration behaviour of our GA. In simulations in the previous chapter, this reduced the overall fitness that could be reached. Therefore, we investigated several modifications of the evolutionary algorithm. The local evaluation of parts of intermediate solutions can significantly speed up the search. In the following sections we present the different approaches and discuss the results.

The idea of breaking up the standard generation and single individual fitness bounds of a evolutionary algorithm is in parts motivated by the area of memetics [Bla99, Daw89, WPG05]. In this context, successful rules can be viewed as something comparable to cultural information that can be transferred between individuals detached from EA generations. We will consider the advantages that might arise with this perspective.

5.4.1 Local Individual- and Generation-Modifications

We presented in the previous sections a successful approach to use cellular automata for learning rules that is unfortunately very slow due to the time consuming calculation of fitness values for intermediate rulesets. Therefore, we investigated several techniques to speed up the fitness development. As a result, we developed an extended GA-operator, the *memetic modification (mm)* operator. The underlying idea is that we rate the single rules of an individual by a kind of additional or internal fitness function.

The Memetic Modification Operator

When we examined the 'successful' rulesets (RSs)[7] we found out that some of the rules will be used in a lot of random cellular automata whereas others will never be used. This means that there are rules which are rather similar or identical to a lot of random assignments in CAs and others that will (almost) never be chosen. We create a list of all rules that have been executed during the evaluation of the set. In the following text, such a set $\mathcal{U}$ is denoted as '(a set of) useful rules'.

The memetic modification operator will now modify an individual $\mathcal{I}$ of the EA (i.e. a set of rules) by replacing all unused rules with randomly chosen rules from $\mathcal{U}$.

If the function $exec(r)$ calculates the number of executions of this rule r on a random CA test setting, and $r_\mathcal{U}$ is at each access a random rule from the useful set of rules $\mathcal{U}$, the mm-operator for an individual $\mathcal{I} = \{r_1, r_2, ..., r_n\}$ is defined by

$$mm(\mathcal{I}) = mm((r_1, r_2, ..., r_n)) = (r_1', r_2', ..., r_n') = \mathcal{I}'$$

with

$$r_i' = \begin{cases} r_i \text{ for } exec(r_i) > 0; \\ r_\mathcal{U} \text{ for } exec(r_i) = 0 \end{cases}$$

In the next part of this section, we describe the embedding of the mm-operator into the existing ones. Thereby, the application of the mm-operator to a population $\mathcal{P}$ will result in the applying of the operator to each individual in $\mathcal{P}$ separately.

Enhancement of the Selection Operator

The selection process of the EA to choose individuals for the next generation will be enhanced by adding the memetic modification. So, the next generation of individuals in the EA will be generated by:

---

[7] these are the sets with a high fitness value

$$G_{t+1} = mm(selection(G_t))$$

We also consider the application of the mm-operator to only a part of the selected individuals from generation $G_t = \{I_1, I_2, ..., I_n\}$:

$$G_{t+1} = (I'_1, I'_2, ..., I'_n) \text{ with } I'_i = \begin{cases} I_i \in G_t \text{ with } prob(p_s); \\ mm(I_i) \qquad\qquad \text{else} \end{cases}$$

One obvious problem with this concept is the assignment of a previous calculated fitness value to a new individual which fitness is yet unknown, because we are changing the rules the individual consists of. Therefore, the selected individuals represent not necessarily fitter individuals, the mm-operator can here be seen as some kind of 'clever' mutation operator. Nevertheless, a standard mutation operator as defined in section 5.3.2 will additionally be assigned to the population with a given probability.

Enhancement of the Crossover Operator

Due to the problems that might occur with the modification of the selection operator mentioned above, we analysed another approach. There, the mm-operator will be applied only to the newly generated children in the population. Hence, we can assure that the selected individuals with high fitness values can remain unchanged in the population. The description of the One-Point crossover operator for two individuals I_1 and I_2 with crossover point p and the new generated children I_3 and I_4 is:

$$cross_{mm}(I_1, I_2, p) = \{I_3, I_4\}$$

with

$$I_1 = (R_1, ..., R_n)$$
$$I_2 = (S_1, ..., S_n)$$
$$I_3 = mm((R_1, ..., R_{p-1}, S_p, ..., S_n))$$
$$I_4 = mm((S_1, ..., S_{p-1}, R_p, ..., R_n))$$

Again, we can add a probability to limit or guide the number of executions of the memetic operator. Then, the first child is produced as follows

$$I_3 = \begin{cases} mm((R_1, ..., R_{p-1}, S_p, ..., S_n)) \text{ with } prob(p_c); \\ ((R_1, ..., R_{p-1}, S_p, ..., S_n)) \qquad\qquad \text{else} \end{cases}$$

and the second one I_4 analogously.

5.4.2 Results

We compared the memetic modifications with each other in several simulation runs
to obtain the best method. In order to achieve this, we considered the modification
of the selection and of the crossover operator separately and compared these results
with the concurrent use of both modified operators as well as with the standard rule
development without the modifications introduced here. The averaged results are visualised in figure 5.12. There is a significant difference between the development of

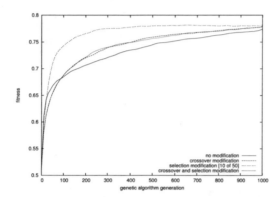

Fig. 5.12. Comparison of the different characteristics of memetic approaches. All curves show the average of
25 learning processes, each required around 8 days of simulation time on a fast machine. The single curves are
smoothed with a Bezier function to enhance readability.

the fitness values in all four settings. After approximately 1000 generations the fitness of all settings is comparable similar, but the developments of the fitness values
differ for each setting. The modification of the selection operator shows the best convergence because the final fitness values region is reached very fast compared to the
other methods. As all methods outperform the standard fitness development, all of the
modifications can enhance the standard learning algorithm.

If we compare the different runs of one method with identical parameters, we can observe that the standard deviation is very low. Hence, we can draw the conclusion that
the different fitness development depends mainly on the used methods. In figure 5.13
the statistical analysis of the most successful modification and the standard approach
is presented. The modification shows in the first approx. 250 generations a slightly
higher value for the standard deviation σ, but from there on this value is below the
deviation of the standard approach. Thus, the convergence behaviour of the particular
runs is very similar.

Now we will examine a selected set of rules. As a comparison of all obtained sets
is an unmanageable approach, we present here only an exemplary sample. However,

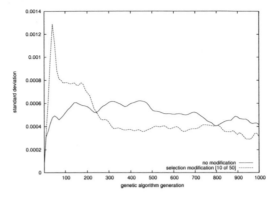

Fig. 5.13. The standard deviation σ of the best modification (the modification of the selection operator) and of the standard approach (solid line).

this can lead to a better understanding of the structure and the connection of a successful set. Therefore, we will visualise in what way the rules in a set depend on each other. In figure 5.14 we use second order statistics to visualise the rule dependencies. A high value on position (i, j) in this graphic denotes a frequent execution of rule j after rule i.

To put it in other words, if for one cell - due to similarity comparisons of the neighbourhood - a rule i is the most appropriate one, and at the next update (next time step) the rule j is the most appropriate one for the same cell, the value in field (i, j) will be increased. We can see that most of the time one rule is repeatedly executed, hence the diagonal is clearly pronounced[8]. But there are some other grid positions with values greater than zero. When we examined such cases, we could find two reasons. Often, the most similar rule for one cell in the CA changes in the first two or three iterations. This was due to changes in the neighbourhood. But after such few steps the neighbourhood remains static and the executed rule remains the same.

The second reason appears less frequent. Rules change alternating from state A to B and back to A again. There can be arbitrary types of states and arbitrary many of them in between. The reason for this behaviour is the interactions with adjacent cells whose states are also changing. This is similar for example to oscillating structures in the Game of Life [Wol02] (so-called *Blinkers*).

As we are interested in the relations of the rules with each other, we arranged the nodes that are involved in the most frequent transitions in a graph structure. The nodes represent the rules that have been executed. A directed edge from node n to n' represents the execution of rule n at time t and the execution of rule n' on the same cell at time $t + 1$. Figure 5.15 depicts such a graph, generated from one set of rules with a high fitness value that has been learned after 1000 generations. Only

---

[8] A position on the diagonal represents the field (i, i), i.e. rule i is consecutively executed.

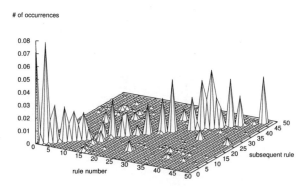

Fig. 5.14. The coherences between two consecutive rule executions. A point at (x, y, z) denotes the fact that - for all cells - the number of executions of rule z after rule x on the same cell occurred in y percent of all transitions.

transitions that occurred often (see figures' 5.15 caption) are plotted to increase readability. One can see that nearly half of the rules have only edges to itself. The others are connected, sometimes in quite large subgraphs. This could be interpreted as long sequences of rule executions, but if one considers the occurring sequences in simulation runs, the length of these sequences is very short. At most sequences with 4 rules could be discovered.

5.4.3 Conclusion

In this section we presented a successful modification of the basic approach to learn effective rules for cellular automata. With all presented modifications we could significantly foster the fitness development in the first generations. With the best one, the modification of the selection operator, a high fitness level can be reached after only approximately 250 generations while the standard approach requires for the same level more than 1000 generations. Hence, the learning process needs only a quarter of the time compared to the standard approach. Additionally, we did some basic analysis of sets of rules that were found with our evolutionary algorithm. As some interim conclusion, we discovered that the dependencies between different rules are not as high as the dependencies of the rules with themselves. Nevertheless, there exist dependencies and it might be interesting to develop an additional modification operator in future research that will not only replace single rules but subgraphs from a rule dependency structure.

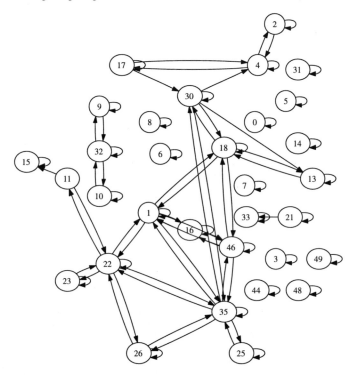

Fig. 5.15. This graph shows the most frequent transitions between the cells. In this example, at least 1000 transitions (from approx 3.000.000) between the same rules have to had occurred to let an edge appear in this graph. This graph is probably not a representative for all such rule structure graphs, but it gives a first impression.

6

Development of Organisations

It is literally true that you can succeed best and quickest
by helping others to succeed

Napoleon Hill

This chapter introduces economic multi agent systems. A project from the computer scientist Michael Schillo [Sch04] will be examined, enhanced, and transformed to heuristics for the *OPP*. For enhancement, we introduce delay, generosity, and locality. In the end, two different approaches will be presented which both are inspired from Schillo's PhD thesis project.

The first of these algorithms permits the agents to choose arbitrary agents to become members in their organisation. This choice is biased by the utility of the additional agents and the supplementary communication costs. The second approach is able to constitute a hierarchical order among the agents. Each agent on each level is only responsible to fulfil orders from one superior agent and to direct a small amount of subordinated agents. Both hierarchy establishment and communication is realised by exclusive local interactions. We will show that the system is highly scalable.

6.1 Economic Multi Agent System

Multi agent systems are used in a wide range of different areas. One of these areas is the modelling of business processes. These highly complex processes can perfectly be represented by a set of agents since economic individuals are acting mostly autonomous and in an egoistic way. One approach has been done in 2003 by Fischer and Schillo [SF03]. There, agents can recursively form unions, so-called *holonic* multi agent systems , that act as a single agent to an outer observer. This chapter uses ideas inspired by Schillo et al. to compare different forms of agent unions as they did in [SFF+04].

We can find other forms of economic multi agent systems in real life very frequently. One example are the bidding agents in the auctioning system eBay[1]. A user informs the agent about the maximum amount of money he or she is willing to pay for an item and the agent will bid until this limit is reached or until there is no need to raise the bidding. Obviously, such agents are autonomous and egoistic. Each relevant action of an agent will be reported to the user.

Another interesting system has been presented by Pattie Maes et al. from the MIT [MC96]. The system[2] does not only provide agents that buy goods, but also selling agents. In the system, all necessary negotiations are included. Each user of the system can create a selling or a buying agent with all the necessary parameters, i.e. the price for the good the agent should pay / achieve. Then, the agents search for an appropriate negotiator. If no buyer or seller could be found, the agents offer reasons why this might has happened.

Hahn [Hah03] extends the approach of Schillo. He distinguishes between a large number[3] of possible economic organisation forms. Each single form is rated with regard to a given economic setting. The best form is searched with the help of a genetic algorithm.

6.2 Schillo's Approach

The idea of Schillo et al. was to design a multi agent system whose internal organisations were based on existing organisation forms in economic science[4]. The aim was to increase the whole productivity of the system. In some organisation forms, such an increase can be achieved due to the decrease of planning activities and the minimisation of communication overhead. A group of producers might be able to combine their single, basic goods to a more complex one. Additionally, he tested a holonic multi agent system with regard to its robustness while processing different online modifications. The aggregation of agents can handle a required action if the jobs are not solvable for single agents. In economic science, the term *coordination costs* describes this elements best.

In their concept, agents are allowed to form organisations, which are agents again. Such agents, he calls them *holonic*[5], are - observed from the outside - not distinguishable from single agents. Hence, several holonic agents can again be combined to another holonic agent and so on, the construction has recursive properties. Since there is no additional object in a holon, at least one of the involved agents has to be a representative for the organisation to the external world. Such an agent is called *head* of an organisation, the non-head agents are called *body* of an organisation. Inside an

---

[1] www.ebay.com

[2] They call this virtual market place 'Kasbah'

[3] in detail, he differentiates between 10368 forms.

[4] He mapped some organisation forms from transportation area

[5] This means 'A Whole which is part of a Whole' and is based on the Greek words *holos*=whole and *on*=to be

organisation, there are bindings or contracts (commitments) between the participating agents which help to keep the organisation together.

This concept of holonic agents has three main advantages:

- The whole system is highly compatible to other multi agent systems since a holonic agent can be addressed like a single agent. For other agents, there is no difference.
- The holonic concept introduces the successful recursion concept into the area of multi agent systems. This can offer new possibilities.
- Since there are no restrictions for the structure of internal organisations inside a holon, the system is easy to extend.

In their FORM[6] system [SFF+04], Schillo et al. distinguish between seven organisation forms. These can be ordered according to the looseness of connections among the agents. Maximum looseness happens between agents that do not interact, minimum looseness between agents that totally give up their individuality. These organisations with maximum and minimum looseness are only theoretical concepts, Schillo et al. concentrate on the 5 forms in between. The amount of looseness in the relations is configured by contracts between the agents.

Next, we present the 7 different forms of organisations, starting with maximum looseness. This is taken from [SFS03]. In general, a higher looseness results in more autonomy for and flexibility of the single agents, but lower looseness can reduce planning and communication expenses [Sch03]. Some of Schillo's experiments show a reduction of messages when transforming to a higher organisation form. At the same time, the overall handled orders[7] stay the same. The whole system is able to adjust itself to a given order structure of the customers.

Single, Autonomous Agent

This is a theoretical concept in which all agents are acting totally autonomous. There is no cooperation or communication between the agents in the multi agent system, the only interaction is between providers and customers.

Market

The market is an extension of the totally autonomous concept. The agents are still acting autonomously, but they are allowed to cooperate. Each agent can forward an order to another agent. This is some kind of reactive behaviour, there is no need that in a similar situation the same agent reacts in a similar way. Hence, there is no need for any kind of representative or head in a market organisation, but the agent that re-delegates the order or parts of the order acts as a holon head for this specific job in this specific situation.

---

[6] A 'Framework for self-Organisation and Robustness in Multiagent systems'
[7] this is one of his quality measurements

Virtual Enterprise

The difference between a market and a virtual enterprise is that agents form a temporary organisation to produce complex goods that none of them could produce individually. But this organisation form is limited to a single order. It is highly important that in such singular fusions the agents can trust each other. We will explain in section 6.3 how this trust can be raised. The head of a virtual enterprise is the agent that initiated the formation of the organisation and delegates the agents. This organisation form offers a highly flexible solution.

Alliance

An alliance is a virtual enterprise that maintains the organisational inter-agent connections after the completion of an order. The representative is non-ambiguous and is determined in an election. The distribution of the money obtained from current and future orders will be defined in advance during establishment of the alliance. The head of an alliance can enhance its economical power by many successful customer contacts, hence it might be useful for an organisation to elect such a powerful agent as its head. Otherwise, the organisation might loose several useful customers.

Strategic Network

The strategic network is - in Schillo's definitions - quite similar to an alliance. The main difference is that the head of the organisation has a higher authority. This leads to a less dynamic structure of the organisation, members stay mostly the same. The single agents offer their internal data to the head of the organisation. The result is, that the head decides what kind of goods each agent should produce. If an agent has contact to a customer, it refuses its call for proposals and redirects the inquiry to the head of the organisation.

Group

In a group, the head of the organisation has very much power. It can decide if an agent is allowed to become a member of the group or not. There is no longer any kind of negotiation, the head decides and body agents are not allowed to call the decisions into question. This authoritarian behaviour will drastically reduce the amount of internal communication since there is no longer any need for offer, negotiation, request, confirmation etc. messages inside an organisation.

Another significant difference is the distribution of the benefit between the agents of the organisation. In this organisation, only the head will obtain the benefit, but it has to regularly pay the agents a salary. This organisation form is the only one that forbids each body agent to be additionally in other organisations, the membership is exclusive.

Corporation

The theoretical organisation form corporation can be seen as an amalgamation of the single agents. The representative of the organisation owns all the knowledge and all the resources from the body agents. This process is not reversible, the agents loose

	Market	Virtual Enterprise	Alliance	Strategic Network	Group
Delegation	Economic	Economic/Gift	Economic/Gift	Authority	Authority
Membership	No limitations	Product spec.	Product spec.	Product spec.[8]	Exclusive
Benefit	Order	Order	Regulation	fixed ratio	regular salary
Head	none	all	one	one	one

Table 6.1. The different forms of organisations and its characteristics. This table is based on the research of Schillo et al. [SFS03]

all their autonomy.

The main characteristics of the different organisation forms are combined in table 6.1.

With such a parameterised multi agent system, Schillo et al. want to increase the robustness of it referred to three different interferences. They consider *dropout, scalability*, and *environmental change*. The results without the interferences in an optimal run will be compared with the results dealing with the interferences. This is done with regard to different aspects. For dropout scenarios, they compare the number of completed orders with the number of orders that could be completed in the optimal run. The reaction of the multi agent system to changes in the environment, i.e. changes in the structure of orders, is measured in the maximum increase of the price, again compared to the optimal runs. The scalability of the system, i.e. the increase of the number of providers and customers, is measured by the amount of additional communication. The results are presented in table 6.2. Two main outcomes are obvious.

	dropout	scalability	environmental change
Market	2.72%	12.09%	0.0%
Virtual Enterprise	2.35%	5.09%	4.4%
Alliance	2.04%	5.16%	4.4%
Strategic Network	2.04%	3.51%	6.45%
Group	1.95%	3.22%	18.25%

Table 6.2. The quality of the organisation forms for handling perturbations. The smaller the values are the better an organisation can adapt to changes. This table is based in information presented in [SFS03]

First, the more restricted the organisation is, the better it can adapt to the failure of agents or to changes in the number of agents. Contrary, the stricter the organisation

---

[8] In the original paper by Schillo [SFS03] he defines in table 2 the membership in a Strategic Network to be 'Product specific'. This is a difference to the description of the organisation form 'Strategic Network'.

is constructed, the harder it is to change internal structures. Hence, such fixed organ-isations can only adjust slowly to changes in the market. If the customers suddenly need other products, it takes time for *groups* to change contracts. On the other side, the loose organisation forms can quickly release contracts, if there are any, and build new organisations in the next round. One consequence from these results is that there is no optimal organisation form, it depends on the expected structure or type of the perturbations whether the agents should have loose or strict connections inside the holon.

6.3 Enhancement of Schillo's Approach

To use Schillo's approach as a solution for the *OPP*, we implemented it according to the descriptions in his PhD thesis and several conference proceedings. But there were still several open points, a lot of ideas and concepts that have not been described in detail with all the necessary parameters etc. During the enhancement, we had several additional ideas and tested them.

6.3.1 Introducing Agent Delay

We examine the influence of an agent specific delay to the trust of other agents, on the one hand within the group of provider agents and on the other hand among providers and customers. The results support the theory about the robustness of the whole multi agent system. Both provider and customer agents are able to figure out which of the agents had high delay values and separate them out. Some results are presented in figure 6.1(a). In these settings, there are $(n \cdot n) - n$ possible agent interaction relations within the group of provider agents. We parameterised half of the agents to have such a delay in varying intensity. Hence, half of the agent interaction relations take place with delayed agents, i.e. $\frac{1}{2} \cdot (n \cdot n) - n$. In the setting visualised in figure 6.1(b) for $n = 10$, the overall number of ignored agents is 44.9[9] and close to the theoretical value of 45. Thus, the economic system is able to ignore the correct number of delayed agents if the delay is high enough. For lower delays, the system is still explicitly better than a system that chooses the interaction partners randomly.

6.3.2 Introducing Agent Generosity

We introduced a generosity factor into the individual agents' parameters to influence the gift exchange between provider agents. A gift exchange is a possibility to modify the trust of other agents in the agent itself. The amount of donated money is set in relation to the wealth of the donor. The higher the relative value of the gift is, the higher the increase of trust will be. This relative value is influenced by the generosity of an agent.

---

[9] which is the average of 20 simulations

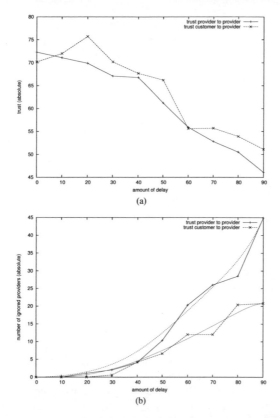

Fig. 6.1. Figure 6.1(a) visualises the overall value of agent trust to all provider agents. Obviously, it decreases when the amount of delay for some provider agents increases. Figure 6.1(b) shows how the provider and the customer agents react, they ignore more agents if the delay for some agents is increased. To enhance the readability, we added an approximation curve to both results.

In our simulations we could not identify a significant influence of this parameter to the overall simulation results. We compared the trust between agents with maximum generosity to the trust between agents with lower generosity, but the maximum difference we could observe was approximately 3.74%, even if we consider a generosity factor of 0% (see figure 6.2). The reason might be that gift exchange only occurs in two organisation forms (Virtual Enterprise and Alliance) and occurs even there quite seldom.

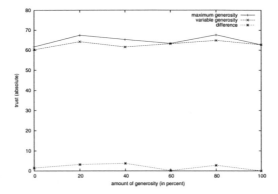

Fig. 6.2. The non-significant differenc of inter-agent trust values for different values of generosity.

6.3.3 Introducing Locality

Next, we add a geometric aspect from the *OPP* to the economic multi agent system by introducing locality into the original approach from Schillo. This is done very straightforward. We allow only agent interactions if both agents are not too far away from each other. We change the maximum interaction distance among agents and compare the outcome of large simulations. The Euclidean distance is calculated both for interactions within the group of providers and among customers and providers. In figure 6.3 the results for very small up to the maximum interaction distances[10] are presented. Again, the results are not very surprising. By an increase of the interaction radius, the number of orders is increasing. This is mainly a result of the additional constraints for stricter organisations to be formed. It can be clearly seen in the development of the number of orders that are handled by organisations. In a limited interaction distance (below 40%) this influences the overall outcome, for larger distances there is no significant decrease. The influence on the orders fullfilled by single agents is only significant for extremely small distances (below 15% of the maximum size). Hence, we can strengthen the results we already obtained in other chapters. If the interaction distance is not too small, the outcome of the local interactions is in random setting not far away from approaches which use global interactions.

6.4 Using Schillo's Approach to solve the *OPP*, Part I

Due to the success of Schillo's approach with the forming of dynamic organisations, we have been inspired to use a similar concept as a heuristic for the *OPP*. The idea is to let the agents decide autonomously if they want to form an organisation. They

---

[10] The maximum interaction distance is the distance where every agent can interact with any other agent in the simulation.

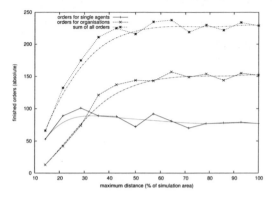

Fig. 6.3. The development of finished orders for different interaction distances and approximation curves for each experiment.

can decide for each additional agent whether it enhances the result if the agent joins the organisation. Again, an agent can be a representative for a whole organisation or group of agents (holonic agents concept). Inside an organisation, we assume that the agents can find an optimal solution for the *OPP* concerning the participating agents. Obviously, if we would consider only this aspect, all agents would form a single organisation and could find the optimal solution. Therefore, we affect the decisions of the agents by introducing communication costs. These costs represent both information exchange inside the organisation and computational power to find the exact solution. Hence, an increase of the number of agents inside an organisation results in an increase of the produced costs. For the consideration of the agent distances, the costs between agents that are located far from each other are higher than the communication costs between nearby located agents[11]. We try to find out how these costs influence the size and number of organisations the agents form. We denote this approach *Organisation Forming Strategy (OFS)*.

6.4.1 The Algorithm

The algorithm that guides each agent and leads to organisations according to the communication cost factor is as follows: We start with the initialisation of the agents with random targets in line 3 of Algorithm 8. If we choose a uniform random number generator, we can assume that after such an initialisation the distribution of the agents onto the different targets is comparatively uniform. Following that, we repeat the loop starting in line 5 until no more changes occur or until a predefined number

---

[11] This is not a substantial difference to the costs we considered earlier. There, we set the number of communication partners to a maximum value with the condition that we have to select the nearest agents. Here, we are allowed to select agents that are not below the nearest ones, but then the costs will be higher. Hence, we again award short and punish high distances.

Algorithm 8 The *Organisation Forming Strategy (OFS)* algorithm as pseudocode

```
 1: procedure MAIN
 2:     for all (agents a in Population) do
 3:         a.currentTarget ← randomTarget
 4:     end for
 5:     repeat
 6:         changed ← FALSE
 7:         for all (agents a in Population) do
 8:             b ← a.findRandomAgent()
 9:             if (a.willAgentImproveOrganisation(b)) then
10:                 a.addAgentToOrganisation(b)
11:                 changed ← TRUE
12:             else
13:                 change nothing
14:             end if
15:         end for
16:     until (changed != TRUE)
17: end procedure
```

of maximum steps has been reached. As a result, each agent will repeatedly do the following steps. First, it chooses a possible new candidate to join its own organisation. This candidate can be an organisation, too. Then, the organisation will calculate an optimal solution for the *OPP* for all participating agents. At the same time, the new organisation costs will be calculated. If the solution quality for the *OPP* instance can be improved and the amount of improvement can compensate the higher organisation costs, the new agent is allowed to join the organisation, otherwise it will not be included. Thus, the organisation form is similar to a strategic network based on Schillo's classification.

6.4.2 Results

Regarding the first modification of economic multi agent systems, several experiments have been made. They all consider the applicability of these ideas to find solutions for the *OPP*. We examine the influence of different communication cost weights on the overall error produced for random *OPP* instances and analyse the arising organisation forms. The rating of the results is no longer depending only on the partitioning and distance objectives, but additionally on the generated communication costs. For that purpose, the evaluation function for an *OPP* solution is expanded to

$$
f = \alpha \cdot \left(\frac{\prod_{i=1}^{m} b_i}{\prod_{i=1}^{m} o_i} \right) + \beta \cdot \left(\frac{\sum_{i=1}^{n} \min_{j=1..m} \left(\delta(a_i, T_j) \right)}{\sum_{i=1}^{n} \delta(a_i, \tau(a_i))} \right) + \gamma \cdot \left(1 - \frac{\sum_{i=1}^{n} \sum_{j=1}^{n} c_{(i,j)} \cdot \delta(a_i, a_j)}{\sum_{i=1}^{n} \sum_{j=1}^{n} \delta(a_i, a_j)} \right)
$$

with $\alpha + \beta + \gamma = 1; \alpha, \beta, \gamma \geq 0$

The new part is the summand with weight γ. $c(i, j)$ is either 1 or 0 depending on the existence of a directed communication link from agent a_i to a_j.

Influence of the Communication Costs

First, we consider the influence of the communication costs onto the overall solution quality for the *OPP*. It is obvious that zero communication costs will result in one large organisation. We assume that such an organisation is able to solve the *OPP* in an optimal way. Hence, zero communication costs can lead to good and sometimes perfect behaviours. But we want to examine how the communication costs influence the behaviour of the OFS strategy. In simulation runs, we found out that the relationship between the communication costs and the error can be approximated by a nearly linear function. This is shown in figure 6.4. Nevertheless, even a maximum consid-

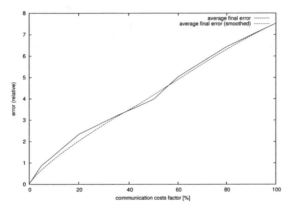

Fig. 6.4. The influence of the communication costs for different cost factors on the final solution quality for the *OPP*.

eration of the communication cost leads to good results, clearly below 8% deviation from the optimum solution in average. Expressed as a linear function, the deviation ν for an arbitrary cost factor x from the optimum can be good approximated by

$$\nu(x) = \frac{3}{40} \cdot x = 0.075x$$

Error Decrease

Now that we have shown that this OFS algorithm is competitive to other strategies, we will show what advantages it has if we compare it with one successful basic strategy (ETS, see section 4.2.3). If we compare the final quality, both strategies are (for given parameters) similar, but we will now concentrate on the error decreasing speed. In figure 6.5 we visualise the level of error over the number of interactions. When nearly no interaction has taken place, the error is very high, due to the bad initial positioning of the agents. But after only approx. 1000 interactions, the error has been halved for

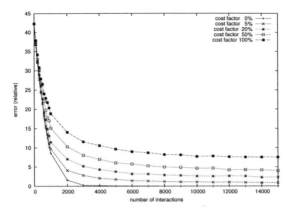

Fig. 6.5. The influence of the communication costs factors on the solution quality for the *OPP*.

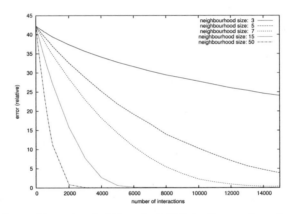

Fig. 6.6. The influence of the number of neighbours for the ETS algorithm according to the error decrease behaviour.

most of the considered cost measured values. When we compare these results with the convergence speed of the ETS algorithm for different neighbourhood sizes as shown in figure 6.6, the differences are striking.

Figure 6.6 shows the same number of interactions as figure 6.5 does, but the error values are clearly different. Only for very large values of the neighbourhood size ($k = 50$) the error decrease is comparatively fast, for smaller values it is much slower. We compare the behaviour of these two algorithms with higher precision in figure 6.7. There, the differences in the error for each number of iterations are depicted. Even the worst performing OFS strategy with the maximum communication costs shows at the beginning a clear advantage compared to the ETS algorithm. It needs several thousand

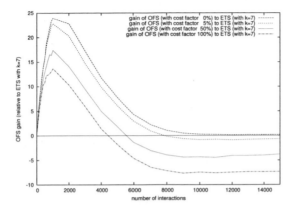

Fig. 6.7. Comparison of the convergence speed for the algorithms OFS and ETS

interactions (here approx. 5000) for the ETS algorithm to have a lower error. However, then it stays better for the remaining number of interactions. But if we consider the quality for low communication costs values, the prior performance is very good and over long time the quality is comparable to ETS.

Comparison of Organisation Sizes

As a last analysis, we will compare the sizes of organisations that appear in the different strategies. Up to now we have not defined the term 'organisation' in the domain of the ETS algorithm (or the k-neighbourhood graphs). For that reason, we denote each agent and its directed neighbours as an organisation. For the OFS strategy we can simply observe the organisations that appear during the simulations. Figure 6.8 presents the results we obtained with groups of 1000 agents. The value in the graph is the maximum number of organisations that appeared at the appropriate communication cost factor. The second function shows the average organisation size. This is low for all communication costs, because a majority of very small organisations and only few but partly large (containing up to 50% of the agents) organisations will be established.

6.5 Using Schillo's Approach to solve the *OPP*, Part II

The first OFS approach introduced an idea allowing the agents to interact with arbitrary other (groups of) holonic agents. In a way, this contradicts the locality characteristics of the considered multi agent systems, even though these communication costs have been taken into account with the modified *OPP* evaluation function. In this section, we present another idea how to use parts of Schillo's approach to find solutions for the *OPP*, the *Hierarchy Forming Strategy*, denoted by *HFS*. We concentrate

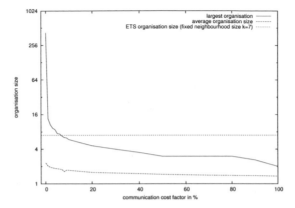

Fig. 6.8. The maximum organisation size that appeared for different communication cost factors. The y-axis is set to logarithmic scale.

on the useful generation of large organisations while focussing on the locality aspect. The restrictions for a fusion of agents to organisations are biased by the topology in the simulation space. The small distances between agents in the same organisation have the advantage that all organisation members probably prefer the same target and that communication costs can be reduced. A similar approach that does not directly consider the locality of the agents but the underlying communication network is presented in [TA04].

In this approach, agents form holons that act as single agents with the same target preference. This process will be repeated several times, until there is only one small organisation with members heading for different targets. On the highest recursion level, one agent is able to analyse the whole situation and to calculate a solution which is close to the optimal solution. There is still a big difference to a global algorithm calculating an optimal solution for all agents 'from the outside'. On the one hand, the formation of the organisations is an emergent process based on only local knowledge and local interactions of the agents. On the other hand, the final calculation of one agent on the highest organisational level does not have to be done for each single agent but only for a very small number of subordinated holons that represent all other agents.

6.5.1 The Algorithm

The main algorithm to distribute the agents in an applicable way - with regard to *OPP* - onto the targets is distributed in two parts. In the first part, the agents form an organisation by recursively combining small groups to holons and choosing one appropriate agent as a holon head. This part ends if the agents cannot find other agents to build a higher level group. The second part performs the actual partitioning of the agents. On each level, the holon heads distribute their subordinated agents according to the knowledge they receive from their holon heads on the higher level.

Organisation Establishment

The algorithm starts with a set of agents that try to form useful groups. These groups can be seen as clusters and the decision for an agent to join a group depends highly on the position of the holon head and the overall size of the group. In the beginning, there is no holon head, thus each agent can collectively decide by communication with each of its k-nearest neighbours which of them - or maybe itself - might be the best candidate for this job. This process would be very communication intensive, therefore the algorithm works in a different way. If each agent sends a notification message ('ping') to its neighbours, a centrally located agent will receive more pings than an agent located at the border of a potential group. This can be motivated by the proof for Theorem 3 and figure 2.7. In each sector around an arbitrary agent a_M there can be at most k communication partners. The more central the agent a_M is located, the more sectors are used by its neighbours[12]. Hence, the in-degree can be augmented up the theoretical maximum of $6 \cdot k$. Due to experimental analysis of random k-neighbourhood graphs, we know that the average in-degree per agent for arbitrary k is significantly below $2 \cdot k$ (see for example figure 2.14). Therefore, we consider the agents with more than k incoming pings to be candidates for holon heads of a group. The better, i.e. more central, the agent is located inside a group, the better it is suited to become the holon's head.

In the second phase, each agent decides for one neighboured holonic head candidate based on the distance to it and the expected size of its group. This is a process that will be repeated several times. If an agent changes its preferred holon head, it has to sign off from the old head. By this procedure it is guaranteed that the resulting organisations have exclusive memberships.

The third phase is a realisation of the recursion in the algorithm. All holon heads try to form groups in the same manner as before, but the neighbourhood changes. An organisation of this level can only be established with other holon heads on the same level. During this higher levelled organisation creation, the holon heads still accept agents from the lower level to join the lower level organisation.

The algorithm terminates if there are less than k agents on the same level. If the holon heads recognise it, the remaining agents build the highest level organisation and the generated holon head on this level is the agent starting and guiding the partitioning process. This behaviour is formalised in Algorithm 9.

Using the notation of Schillo, we can describe the in this way established organisations to have an *election* system for the *single* head. The membership is *exclusive* and the aim of the organisation is *the fulfilment of an OPP task*. After this job, there is *no need for the organisation to be continued*. The holon head is the *only decision maker*

---

[12] When we speak about 'neighbours' in remaining part of this section, we mean the k nearest neighbours.

that constitutes whether an agent is allowed to join the organisation. If the agents have

---

Algorithm 9 The generation of holonic groups. This algorithm is executed by each agent in parallel.

```
 1: procedure AGENTACTION(level)
 2:     finished ← FALSE
 3:     repeat
 4:         calculateKNeighbourhood(level)
 5:         if (no neighbour found) then
 6:             globalHead ← TRUE                          ▷ The agent is the head for all other agents
 7:             finished ← TRUE
 8:         else
 9:             sendPingToAllNearestNeighbours()
10:         end if
11:         if (numberOfReceivedPings > k) then
12:             offer all neighboured agents to become holon head
13:         else
14:             wait for neighbours offering to be head
15:             choose the best promoting agent to be head
16:         end if
17:         if (I am head) then
18:             agentAction(level + 1)
19:         else
20:             store holon head
21:             finished ← TRUE
22:         end if
23:     until (finished = TRUE)
24: end procedure
```

---

several possible holon heads which offer themselves to become head of an organisation, the agents choose the best one. (Algorithm 9, line 15). This is quite informally expressed. In our implementation, three rating criteria are considered: The distance to the possible head, the number of pings, and the number of votes this agent received will be normalised and summed up. The holon with the highest rate will be chosen by the agent to become its head. In the next decision round, the head can be changed again. If there are holon heads with very few subordinated agents (we chose a threshold of $\frac{k}{2}$), these heads avoid maintaining the holon head status and all agents that are involved in the organisation voting process looking for a more promising head.

Partitioning of the Agents

The partitioning process is again a recursive process. First of all, all holon heads ask all agents in their organisation for their target preference. This number is published to the holon head on the next (higher) level. At the end, the head on the highest level has a detailed picture of the current situation. Then, the recursion starts. On each level, starting from the highest one, the head distributes the agents onto the targets according to the optimal Algorithm 5 presented in section 3.2.2. The distance difference value in the algorithm is the distance difference value from the holon head on the lower level. This value is representative for the distance values of all agents in this organisation[13]. During the execution of the algorithm, the organisation sizes will be taken into account. If the holon head is deciding that each agent in a subordinated holon should - as an entire group - choose the same target, the corresponding head will give this order to all of its subordinated agents in the next recursion step. If the holon head decides that a part of the agents of a single organisation have to choose target T_1 and another part T_2, this order will be given to the appropriate subordinated head and this agent will execute the order in the next recursion level.

Algorithm 10 The partitioning of the organisations, initiated by the highest level holon head

```
 1: procedure COUNTAGENTS(())
 2:     if (I am global head) then
 3:         send counting order to all members of my organisation
 4:         sum up answers
 5:     end if
 6:     if (I received counting order) then
 7:         if (I am holon head) then
 8:             send counting order to all members of my organisation
 9:             sum up answers and send to my holon head
10:         else
11:             send target preference to my head
12:         end if
13:     end if
14: end procedure

15: procedure PARTITIONAGENTS(toTarget1, toTarget2)
16:     if (toTarget1 == NULL AND toTarget2 == NULL) then
17:         countAgents()
18:     else                                                ▷ here we use Algorithm 5
19:         calculate optimal partitioning for my organisation according to the
20:         distribution percentage toTarget1 and toTarget2
21:         for all agents a in neighbourhood do
22:             if (subordinated agent should go to target T₁) then
23:                 a.partitionAgents(100, 0)
24:             end if
25:             if (subordinated agent should go to target T₂) then
26:                 a.partitionAgents(0, 100)
27:             end if
28:             if (subordinated agent should send x% of organisation to T₁) then
29:                 a.partitionAgents(x, 100 − x)
30:             end if
31:         end for
32:     end if
33: end procedure
```

---

[13] This is a valid assumption since the holon head has been chosen as a good representative for the group due to its local centrality.

6.5.2 Results

We used two test scenarios to rate this *HCS* heuristic. The first scenario is a random distribution of agents and targets in space. The second one considers an accumulation of the agents in a limited area, the targets are again distributed in the whole space. For the agent accumulation, we used a Gaussian distribution with centre in the upper left corner. Such an scenario is presented in figure 6.9.

Fig. 6.9. Example scenario with agent accumulation in one area

Local Connectivity of Organisation Members

It is hard to formalise the quality of the arising organisational structures, therefore, we present in figure 6.10 a typical outcome. We coloured the holon heads with different colours, depending on their level. Two results can easily be reviewed interpreting this figure. First, the holon heads are located central in their individual organisations. Second, agents that are in a common organisation are located nearby each other. With this result, two main goals have been achieved.

Another result from our simulations is the small number of levels. Due to the construction of the organisations, there can be at most $O(\log_k n)$ levels, but the simulations show that at most $\frac{\log_k n}{2}$ levels appear in all the random scenarios. This is a nice result since each level produces additional messages and the establishment of the whole tree-like structure needs more time for any additional level. One additional characteristic of the organisational structure is that it can be drawn as a forest containing only balanced trees. In all simulation runs, the forest contained only one single tree, but it is possible that there appear several of these trees.

Influence of Different k-neighbourhood Sizes

To rate the quality of the usage of organisations, we compare the outcome of different neighbourhood sizes with the simple strategy choosing the nearest target. This strategy has been chosen for comparison since the presented approach with $k = 0$ shows the same behaviour (if there is no neighbourhood, each agent / holon will choose the nearest target). The simulations have been set up for the random and the accumulated agent setting, the results are average values obtained from 50 runs. The introduced approaches show a significantly better behaviour compared to the basic NTS algorithm.

The difference is particularly clear in the accumulated setting (see figure 6.12). But even in the random setting, the error can be at least halved for very small k (see figure 6.11). The higher the neighbourhood size is chosen, the lower the error value is. For large neighbourhoods the error comes close to 0%, but even for $k = 2$, the error is clearly below 3% and therewith a quarter of the NTS error. In the same figure the fitness that could be reached with the ETS strategy (see section 4.2.3) is displayed again for comparison reasons. For small neighbourhood sizes, the HFS algorithm has a clear advantage compared to ETS. But for larger neighbourhood sizes, the ETS can slighly outperform this approach. In the accumulated setting, the choice of k has not a great influence on the error value, even $k = 1$ has an error close to 2% and this is far away from the NTS error with more than 26%.

Considering the communication costs, we have to examine the communication overhead during the construction of the organisation graph. In the next section, we will show that this overhead is marginal for small values of k. And since we have seen in this section that such small k values produce very small errors, the hierarchical approach is a promising heuristic for the *OPP*.

Communication Overhead

In Schillo's original approach, one goal for introducing the holonic organisations was the minimisation of inter-agent messages. Due to the fact that this approach is inspired

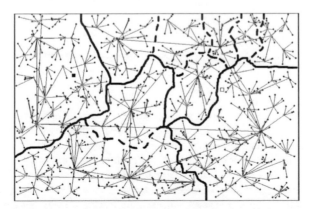

Fig. 6.10. An example of a locally developed global hierarchy. The different colours represent the levels of the organisation heads, the thick lines help to clarify the organisation's border. The squares (black and white) are the targets in this setting. Head of the whole group of agents is the red dotted agent, members of its organisation are the pink agents. (Inspired by [Ken06])

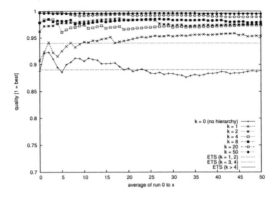

Fig. 6.11. Comparison of neighbourhood sizes in the random scenario. In this setting, we distributed 1000 agents on 2 targets.

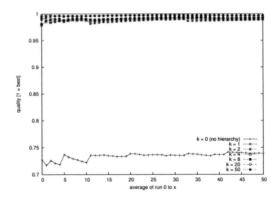

Fig. 6.12. Comparison of neighbourhood sizes in the accumulated scenario. In this setting, we distributed 1000 agents on 2 targets.

by Schillo's idea, the messages between agents to find and spread appropriate *OPP* solutions will be examined. From the experiments depicted in figure 6.13, several conclusions can be drawn. There is a close to linear dependency between the size of the neighbourhood k and the average number of messages per agent. This is traceable when we consider the algorithm in detail. For each agent, there is a minimum number of five[14] messages that have to be done for joining and being in an organisation. Additionally, there are k ping messages and the messages of each holon head. Since the number of holon heads per level is decreasing logarithmically with $\log_k n$, these latter messages do not have a large influence on the overall number of messages. In contrast, the dependency between the overall number of agents and the average

---

[14] *vote*-message, *hired*-message, *count*-command, *count*-answer, and *goto target*-message

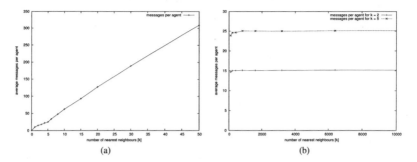

Fig. 6.13. The average number of messages per agent. Figure 6.13(a) shows the approximately linear increase of the number of messages depending on k. Figure 6.13(b) suggests that the influence of the overall number of agents is quasi independent from the average number of inter-agent messages.

number of messages per agent is quite different. There is only a very slow increase observable. Considering agent numbers n in the interval $[100; 10000]$ the average message increase from the smallest to the largest n is below 5%. The main reason for this (in the face of scalability) positive characteristic is the predominantly local behaviour of the algorithm.

6.6 Conclusion

The usefulness of economic inspired organisations based on a model of Schillo has been examined in this section. First, his approach has been extended by some additional agent characteristics. Delay, generosity, and locality have been introduced. In the sections 6.4 and 6.5, these ideas have been used to find solutions for *OPP* instances. Both approaches are quite different. In the first one, the agents form arbitrary organisations which are limited solely by incorporating communication cost. We assume that a single organisation is able to find an optimal *OPP* solution. With these ideas, a very fast convergence to a good solution could be reached. Thereby, the number and size of organisations depend highly on the communication cost weight.

The second approach is able to create a global organisation based only on local decisions and interactions. Such an organisation is able to find good solutions for the *OPP* while it is highly scalable. As a difference to the other approaches presented up to now, these algorithms are able to cope with such graphs which would not be connected when using the basic k-neighbourhood graph structure.

7

Learning Communication

*The people who talk the best are not the only ones
who can tell you the most interesting things.*

Chinese Proverb

Coordination of altruistic agents to solve optimisation problems can be significantly enhanced when inter-agent communication is allowed. In this chapter we present an evolutionary approach to learn optimal communication structures for large groups of agents. The agents learn to solve the *Online Partitioning Problem*, but our ideas can easily be adapted to other problem fields where an efficient communication structure among spatially distributed agents is needed. With our approach we can find the optimal communication partners for each agent in a static environment. In an unknown environment we will figure out a simple relation between each position of agents in space and the optimal number of communication partners. A concept for the establishment of relevant communication connections between selected agents will be provided whereby the space the agents are located in will be divided into several regions. These regions will be described mathematically. After a learning process the algorithm assigns an appropriate number of communication partners for every agent in an - arbitrary located - group.
Parts of this chapter have been published in [Goe06a].

7.1 Introduction

This chapter presents two different ideas in the first two sections. In the static approach, we try to learn one *specific* communication structure for a given and *fixed* setting. With the help of an evolutionary algorithm, each agent finds its optimal communication partners. The whole structure is rated by a fitness function considering solutions for the *OPP* and the occurring communication costs. We are able to find a near optimal communication structure very fast.

The second, 'dynamic' approach differs substantially from the static one. We cannot learn the best communication partners for each arbitrary agent setting[1] by considering only one fixed setting. Thus, we have to find communication characteristics for special regions in the simulation space and apply these characteristics to the agents when they compute to be inside such a region.

A directed communication connection from agent a to agent b means that agent a can ask agent b for information regarding its current target decision. Depending on the setting, the same communication connection can give agent a authority to issue directives to agent b.

7.2 The Static Approach

In this section we consider a static setting, i.e. a set of agents on fixed positions in a two-dimensional, Euclidean space dealing with the *OPP*. The agents have to decide which target to aim for, with regard to the objectives we mentioned in the problem introduction chapter 3.1. For given parameters, which dictate communication costs and parametrise the objective function for the OPP, we try to find an optimal communication structure among the agents. Because of the unknown structure and the size of the solution space we apply an evolutionary algorithm to search for good solutions. With this approach, a solution that optimally fulfils the evaluation function can be found very fast. We will not describe this algorithm in detail. An extensive description and the results can be found in a master thesis [Bec05] that was done under our supervision.

7.2.1 Evaluate the Quality of a Communication Structure

The quality or fitness f of a communication structure can be calculated at any time by the following equation. We sum up the single optimisation criteria, i.e. the partitioning quality, the distance quality and the communication costs, and weight the single parts.

$$f = \alpha \cdot \left(\frac{\prod_{i=1}^{m} b_i}{\prod_{i=1}^{m} o_i} \right) + \beta \cdot \left(\frac{\sum_{i=1}^{n} \min_{j=1..m} (\delta(a_i, T_j))}{\sum_{i=1}^{n} \delta(a_i, \tau(a_i))} \right) + \gamma \cdot \left(1 - \frac{\sum_{i=1}^{n} \sum_{j=1}^{n} c_{(i,j)} \cdot \delta(a_i, a_j)}{\sum_{i=1}^{n} \sum_{j=1}^{n} \delta(a_i, a_j)} \right)$$

with $\alpha + \beta + \gamma = 1; \alpha, \beta, \gamma \geq 0$

$c(i, j)$ is either 1 or 0 depending on the existence of a directed communication link from agent a_i to a_j. The highest possible fitness is $f = 1.0$. The first part is the fitness

---

[1] by 'arbitrary agent settings' we mean settings that differ in agent quantity and agent positions. This unawareness of the environment is reflected by the term 'dynamic'.

function for the initial OPP, as already presented[2] in equation 3.1. This has been extended by the consideration of the fitness costs in the third summand, weighted by γ. We use the same equation for the rating of organisation structures in section 6.4.2. This function incorporates the number of connections as well as the communication distances. The longer the communication distances are and the more communication partners will be chosen, the worse the communication fitness of the whole system is.

7.2.2 The Evolutionary Algorithm

We implemented a standard evolutionary algorithm and guide the search among all possible communication structures by the mentioned fitness function that consists of the three summands representing the different objectives.

Individual

One individual in our GA is a $(n \times n)$-matrix $\mathcal{C}$ describing the connectivity of the agents among each other. A '1' on position (i, j) allows agent a_i to communicate with agent a_j (directed communication). In other words, $\mathcal{C}$ is the adjacency matrix of

$\mathcal{C}$	Agent 1	Agent 2	Agent 3	...	Agent n
Agent 1	0	0	1	...	1
Agent 2	1	0	0	...	1
Agent 3	1	0	0	...	1
...	...	...	...	...	...
Agent n	0	1	1	...	0

the communication graph of the agents.

A complete algorithm would search through all possible table assignments and rate these ones. This is not acceptably feasible since there are 2^{n^2-n} possible assignments[3].

Operators

As a selection operator we use the *Best* selection and for mutation we simply swap bits in $\mathcal{C}$ with a low probability. The crossover method is a modification of the *Single-Point Crossover*. We apply this operator to two communication matrices $\mathcal{C}_1$ and $\mathcal{C}_2$ by choosing a random field (i, j) with $i, j \in \{1, ..., n\}$ in the communication matrix. The two new individuals will exchange a corresponding rectangular part of the matrix defined by (i, j) as the upper left and (n, n) as the lower right corner. We will describe this operator in a more formalised way. Two new individuals $\mathcal{C}_1'$ and $\mathcal{C}_2'$ (children) will be created according to:

---

[2] Of course, in this version, α and β do not longer sum up to 1.0 but to $1.0 - \gamma$

[3] We assume that a connection from the agent to itself is not useful since in our model the agent has access to its own information without raising the communication costs.

$$\forall 1 \leq k, l \leq n : \mathcal{U}(C)_{(k,l)} = \begin{cases} C_{(k,l)}; \forall (k,l) \text{ with } (k \geq i) \wedge (l \geq j); \\ 0 \hspace{4.5cm} else \end{cases}$$

$$\forall 1 \leq k, l \leq n : \mathcal{L}(C)_{(k,l)} = \begin{cases} 0; \hspace{0.5cm} \forall (k,l) \text{ with } (k \geq i) \wedge (l \geq j); \\ C_{(k,l)} \hspace{4cm} else \end{cases}$$

$$C'_1 = \mathcal{U}(C_1) + \mathcal{L}(C_2); \; C'_2 = \mathcal{U}(C_2) + \mathcal{L}(C_1)$$

With this proceeding for the crossover operator we hope to maintain useful communication substructures in the table. This corresponds to the *Building Block Hypothesis*. A detailed description of this approach and its expected advantages in several areas can be found for example in [Hol75] or [Gol89]. There, it has been rated very successfully.

7.2.3 Results

One problem with this approach is the size of the communication table. It grows quadratically with the number of agents. We applied this algorithm several times to a large number of agents (several hundred), but the convergence of our evolutionary algorithm was very slow. A small example for a successful run is presented in figure 7.1. There, after several generations, a good solution that is close to the optimal one could be found. In this picture, the development from an initial random communication structure to a very sparse one can be observed. At the same time, the distribution quality of the agents onto the targets could be enhanced. Hence, only such agents communicate, whose information exchange can enhance the *OPP* solution quality.

7.3 The Dynamic Approach

In the prior section we introduced a static approach to learn optimal communication structures for a given set of fixed agents. But this structure strongly depends on the special setting it was trained on and cannot be used for other groups of agents or other positions of the same agents, especially if we consider communication costs that are constrained by the distance between two communication partners. In this section, we present an approach for agents to learn a useful communication structure which is independent from the distribution of the agents in space.

In order to achieve this, we construct a communication network for a dynamic setting in a different way. The agents do not learn *what* the best communication partners are, but they try to find out *how many* communication partners are useful for the position they are located at (depending on the position of the targets). That means, the agents learn a function that connects a region that can be computed locally with an ideal number of communication partners. Since the agents do not know the extension of the simulation area, the agents have to calculate the area they are in only with regard

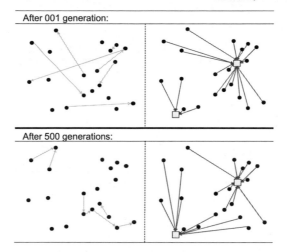

Fig. 7.1. The upper figures show the communication network (left) and the distribution of the agents onto the targets (right) after the first generation of the evolutionary algorithm. The communication connections are randomly distributed among the agents and the target decision can be described as the result of a simple nearest neighbour selection algorithm. After several generations of the GA (here, we show the situation after 500 generations), the communication network is very sparse but seems to be useful since the resulting distribution is very close to the optimal one. This presented distribution has a fitness of 0.989.

to the distances to the targets. Therefore, we develop a so-called q-value for each agent position (x, y) by

$$q_{x,y} = \frac{\min(\delta((x,y), T_1), ..., \delta((x,y), T_n))}{\max(\delta((x,y), T_1), ..., \delta((x,y), T_n))} \in \mathbb{R} \qquad (7.1)$$

Or, to put it more informally, we calculate the quotient for each agent position by dividing the distance to the closest target by the distance to the farthest target[4]. With this procedure we can determine relative values that represent the area an agent is in without paying direct attention to its absolute distance.

Because we are in a continuous space, there is an infinite number of different q-values. Therefore, we combine the q-values in intervals of the same size. For l categories, we obtain the following intervals $I_1, ..., I_l$ with

$$\forall k = 1...l : I_k = \begin{cases} \left[\frac{k-1}{l}; \frac{k}{l}\right[& for\ k \neq l \\ \left[\frac{k-1}{l}; 1\right] & for\ k = l \end{cases}$$

---

[4] In this section we focus on settings with two targets. If we consider a higher number, we will possibly have to make the calculation of the q-value more complicated.

7.3.1 Our Approach

We solve the *OPP* for large agent sets and enhance the knowledge base of a selection of agents by enabling communication with its neighbours. Therefore, the agents have

q-interval	I_1	I_2	...	I_l
number of communication partners	n_1	n_2	...	n_l

Table 7.1. The number of communication partners based on the q-interval. For each interval, an appropriate number can be defined.

to learn the number of communication partners depending on the q-interval they are located in. Or, in other words, they learn an appropriate assignment for each n_i in table 7.1. We call such a table q-table.

7.3.2 Size and Characteristics of the q-Intervals in Space

In this section we will provide a short mathematical insight into the regions we created with our intervals. We consider an arbitrary q-value, denoted by q'. All positions in space that produce exactly the value q' are located on two disjoint circles with the same radius r around two centre points. The targets are inside this circles. The Euclidean distance between the two targets is arbitrary but fixed and denoted by D.

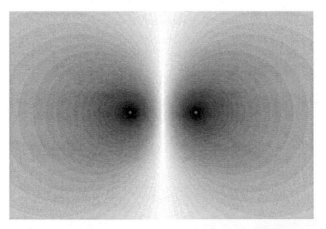

Fig. 7.2. This Figure shows the distribution of the q-values when calculated for arbitrary positions in space. Each grey tone represents one specific q-interval. In this example there are 30 different q-intervals.

Theorem 13. *All points in space that have one specific q-value according to two targets T_1 and T_2 on positions (T_{1_x}, T_{1_y}) and (T_{2_x}, T_{2_y}) lie on the circles C_1 and C_2 with centre points*

$$M_1 = \left(\left(\frac{T_{2_x} - q^2 \cdot T_{1_x}}{1 - q^2} \right), \left(\frac{T_{2_y} - q^2 \cdot T_{1_y}}{1 - q^2} \right) \right)$$

$$M_2 = \left(\left(\frac{T_{1_x} - q^2 \cdot T_{2_x}}{1 - q^2} \right), \left(\frac{T_{1_y} - q^2 \cdot T_{2_y}}{1 - q^2} \right) \right)$$

and radius

$$r = \frac{q \cdot D}{(1 - q^2)}$$

The proof for this Theorem can be found in the appendix A.2. The circles whose properties we calculated in the proof are visualised in figure 7.3.

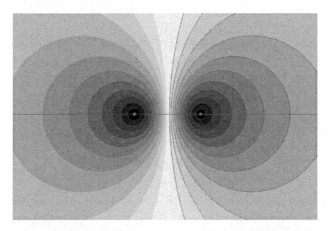

Fig. 7.3. In this figure the circles for the interval borders (here we have 12 intervals), obtained by the mathematical examination, are visualised on the right side. They perfectly cover the regions.

7.3.3 The Evolutionary Algorithm

There is an extensive number of possible assignments for such a structure as presented in table 7.1, especially when dealing with large numbers of agents. Each n_i can be assigned to a value from $\{0, ..., (n-1)\}$ with $n = |\mathcal{A}|$ representing the number of agents. Therefore, we have $(n-1)^l$ possible assignments. We use an evolutionary algorithm to search for good ones because we have no prior information about the structure of the solution space.

Individual

One individual in the evolutionary algorithm represents a table assignment. Each field in the table can have values from 0 to $n-1$.

Fitness

The fitness f of a solution obtained with a table assignment is the quality of the solution of the *OPP*. It is set in relation to the optimal solution regarding to communication costs, and is calculated by the equation:

$$f = \alpha \cdot \left(\frac{\prod\limits_{i=1}^{m} b_i}{\prod\limits_{i=1}^{m} o_i} \right) + \beta \cdot \left(\frac{\sum\limits_{i=1}^{n} \min\limits_{j=1..m} (\delta(a_i, T_j))}{\sum\limits_{i=1}^{n} \delta(a_i, \tau(a_i))} \right) + \gamma \cdot fitness_{Com}$$

$$\text{with } \alpha + \beta + \gamma = 1; \alpha, \beta, \gamma \geq 0$$

We use the same notation as defined in section 7.2, only the communication costs part is different.

The **Communication Costs** in the fitness function will be calculated by taking the communication distances between two agents that are allowed to communicate into account. The number of communication partners is defined in the q-table. When establishing n_k connections for an agent located in the interval I_k, this agent will create communication lines to its n_k nearest agents. The sum of these costs will then be set in relation to the maximum costs of communication that could appear if it holds for each entry in the q-table that n_i is equal to $(n - 1)$. Hence, we can define a partial fitness function of the communication costs:

$$fitness_{Com} = \frac{q\text{-table defined communication graph costs}}{\text{complete communication graph costs}}$$

We assume that a subset of agents that own communication connections among each other will be able to calculate an optimal partial solution for the *OPP*.

Operators

Selection

For the selection of individuals for the next generation we implemented the *Roulette Wheel* and the *Best* selection. In comparison runs the *Best* selection shows slightly better results; therefore, we conducted our final experiments with this method. For both algorithms, we use a (μ, λ)-selection scheme and chose 50% of the individuals for the new generation out of the old one.

Crossover

As a crossover operator we use *One-Point Crossover*, due to its easy application. To achieve this, a random point p from $\{1, ..., l\}$ is chosen. Then we create two new q-tables by recombining the tables from two parental individuals split at this particular point (or column) p. If coherences between the table entries exist, we can manage to maintain them with this operator (see 'Building Block Hypthesis', earlier in this chapter).

Mutation

For the mutation of a q-table we make use of two mutation parameters. p_i defines the probability for mutating one individual. The second parameter p_t determines the probability of mutating one table entry.

If an entry I_k has been selected for mutation, we adjust the value n_k in the table by adding a random value $r \in \{-(n - 1), ..., (n - 1)\}$ to the entry and control if the value is out of range with the equation

$$mutation(n_k) = max(min(n - 1, n_k + r), 0)$$

7.3.4 Our Algorithm

We use an evolutionary algorithm to find the appropriate number of communication partners for each interval in the q-table. The most interesting part of the algorithm is the calculation of the fitness of an individual in the population, i.e. the fitness of a q-table. To rate such a q-table Q, we test the quality of a set of agents working on the *Online Partitioning Problem* that use a communication structure developed from Q. The fitness of each Q can be calculated by the following Algorithm 11.

The function $calculateCommunicationFitness(agentSet)$ simply applies the communication fitness function as described in 7.3.3. The higher this value is the less

---

Algorithm 11 The overall fitness calculation

```
1: procedure CALCULATEFITNESS(qTable Q)
2:     randomSet ← new random set of agents
3:     place targets on random positions in space
4:     agentSet.createCommunicationGraph(Q)
5:     commFitness ← calculateCommunicationFitness(agentSet)
6:     oppFitness ← calculateOPPSolutionFitness(agentSet)
7:                                                          ▷ see Algorithm 12
8:     return α · oppFitness + (1 − α) · commFitness
9: end procedure
```

---

communication is used.[5] The more interesting function is the one calculating the *OPP* solution fitness, as shown in Algorithm 12.

The function $calculateLocalOptimalSolution(a, communicationPartners)$ in line 6 assumes that an agent can calculate the optimal partitioning for the agents it communicates with. This is quite an idealistic picture because we still have a hard problem, but we can take this calculation power into account by increasing the influence of the communication costs for the fitness function. Even if this function can calculate only an approximation, our algorithm still works fine[6].

---

[5] In our simulations we repeated lines 2-6 several (five) times to obtain more meaningful fitness values.

[6] The pseudocode algorithms show only the most important steps of our algorithm, for a more detailed insight we refer to the original Java sources that are available for download and further experiments on our webpage (http://www.upb.de/cs/ag-klbue/de/staff/agoebels/index.html)

Algorithm 12 The *OPP* fitness calculation

 1: **procedure** CALCULATEOPPSOLUTIONFITNESS(AgentSet agentSet)
 2:    $d \leftarrow$ minimal overall distance to targets in optimal partitioning   ▷ reference value
 3:    $s \leftarrow$ optimal number of agents on each target in optimal partitioning
 4:    **for all** agents a in agentSet **do**
 5:       **if** (#outgoingConnections(a) > 0) **then**   ▷ derived from qTable
 6:          $calculateLocalOptimalSolution(a, communicationPartners)$
 7:       **else**
 8:          choose nearest target
 9:       **end if**
10:    **end for**
11:    $d' \leftarrow calculateDistanceFitness(agentSet, d)$
12:    $s' \leftarrow calculateDistributionFitness(agentSet, s)$
13:    return $\beta \cdot c' + (1 - \beta) \cdot s'$
14: **end procedure**

7.4 Results

In this chapter we focus on the results for the dynamic approach presented in section 7.3, whereas more results for the static approach (section 7.2) are presented in [Bec05]. There, a good communication matrix could be found fast for every given fixed set of agents.

7.4.1 Parameters

To obtain meaningful results and to reduce the outlier percentage, each of the simulation runs have been made 25 times. We have chosen the number of q-table entries from the interval $[1, 20]$ and the number of agents from the interval $[25, 1000]$. In the remaining part we present results for the fitness function weights $\alpha = \beta = \gamma = \frac{1}{3}$. We allow the evolutionary algorithm to learn good tables during up to 5000 generations with a population size of 100 and a mutation probability of $p_i = 0.05 = p_t$.

7.4.2 Fitness Development

First of all, we examined how the fitness of the GA is developing. Figure 7.4 shows a typical fitness development. Both the best and the average fitness rise very fast to a high level and remain there. As a reference we present the fitness value of a non-communicative algorithm choosing always the nearest target for each agent (NBS, see 4.1.3). We are restricted to such basic algorithms since we want to set our approach in relation to an approach that does not use communication. The reference fitness is significantly lower than the fitness value achieved with our approach. By adjusting the weights for the communication costs we can obtain any fitness value between 1.0 (No communication cost when γ is chosen to be zero. Then, global communication is possible without any drawback on the solution fitness) and the reference function ($\gamma = 1$, only communication costs will be regarded without any consideration for the other objectives). Hence, we can conclude that inter-agent communication enhances

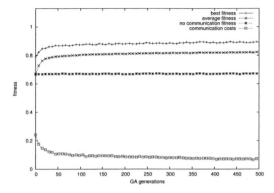

Fig. 7.4. This figure presents the development of the fitness over 500 generations. We show the average fitness of the whole generation and the fitness development of the best individual in population. This graph illustrates the average result over 25 runs. The parameters we used can be found in the source code package. The fitness rises while the communication costs can be reduced.

the solution quality of the whole group and our approach achieves good *OPP* solutions for given communication costs or restrictions.

A comparison between the influence of the agent number and the different upper bounds for the number of communication partners is depicted in figure 7.5. For bet-

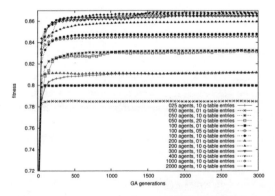

Fig. 7.5. The average fitness development of 25 runs each for different settings (selection).

ter readability, the final results from these simulations are listed in table C.1 in the appendix. Several conclusions can be derived from these values, we arrive at the following points:

1. A number of q-table entries that is too small results in lower fitness values
2. The number of agents influences the maximum fitness that can be reached

3. If a certain number of agents is reached, no significant differences in the maximum fitness value that is reached can be observed

The last two points will be considered in more detail in figure 7.6. As a consequence

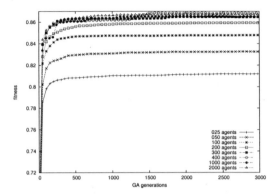

Fig. 7.6. A comparison of the fitness development in settings with different numbers of agents. We changed the number of agents and compared the average fitness in 25 runs of the evolutionary algorithm with 3000 generations each. The number of q-table entries is fixed to 10.

from these points, we can conclude that the approach can best be applied to large agent groups. If there are too few, the learned solution might be worse.

7.4.3 Limited q-Values

We continue our research by examining the performance of our algorithm when we limit the maximum number of communication partners per agent. Therefore, we observe the fitness development if we apply different upper bounds to the values in the q-table. These restrictions are useful in real systems where communication costs occur. An unlimited number of communication partners (and with it an unlimited number of messages) is not realistic with regard to costs, physical properties, and energy consumption. The results are visualised in figure 7.7. As one might expect, the upper bounds seem to have a direct and significant influence onto the maximum fitness that could be reached. But the evolutionary algorithm is still able to find good solutions.

7.4.4 Development of q-Table Values

As our approach seems to be applicable, we want to gain a better insight into the communication structure learned by the agents. We approach this topic by observing the changes of the q-table entries during the learning process. Figure 7.8 displays a typical picture representing the development of the values. After several hundred

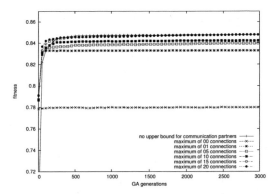

Fig. 7.7. The different fitness values that appear if we add upper bounds for the q-table assignments. The best fitness can be reached if there are no upper bounds. The worst one appears if we allow no communication (upper bound of 0). This example is taken from a simulation run with 100 agents. The maximum assignment learned in an unbounded run is 19. Hence, an upper bound above 20 does not make sense for the overall fitness, but the convergence can be enhanced. The single graphs are average developments from 25 simulation runs.

communication partners

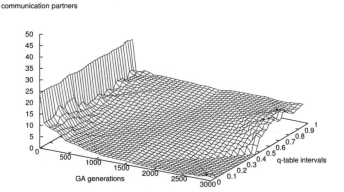

Fig. 7.8. The development of the q-table entries during the learning process.

generations the q-tables show similar results for all settings. In this figure the average over all q-tables in 25 runs is presented. In the early generations the number of communication partners is high and nearly identical for all intervals. But the communication structure becomes sparser fast and in the later generations we can see that the q-table can be divided into 2 parts. For intervals containing small q-values the agents learn to have nearly no communication partners[7]. For q-values greater than 0.4 it seems to make sense to communicate with a small number of neighbours to increase the fitness to a near optimum value. One other key result is that the overall number of communication connections between agents is very low compared to the maximum possible value. This leads us to the conclusion that a communication structure does not necessarily have to be very complex or dense to be effective if we generate it in an intelligent way.

7.4.5 Calculation of q-Table Entries

Up to this point of our research, the values for the q-tables have been discovered by an evolutionary algorithm. For large settings, this can be a very time consuming procedure. It would be more effective to have a function that can fill the q-table for an arbitrary agent number with appropriate values. Therefore, we first consider the optimal q-table entries deducted in several experiments. Some examples are visible straight away in figure 7.9. It is obvious that the number of communication partners

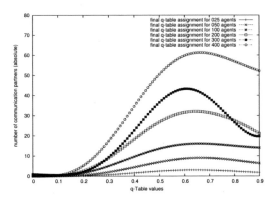

Fig. 7.9. This figure shows the absolute number of communication partners for each q-table entry. These numbers depend on the number of agents in the simulation.

strongly depend on the number of agents in the simulation. In this figure, the differences for the single q-table entries can be seen clearly. Another point that can be discovered is the similar characteristic of each of the curves.

If the values are translated into relative values, the differences between the curves become less significant. Therefore, we calculate the percentage value for each table entry in relation to the overall number of agents. Figure 7.10 shows the result of this

---

[7] The intervals I_1 and I_2 contain all q-values below 0.1. As we saw in 7.3.2, the area in space representing all possible values in these intervals is very small compared to the remaining space, hence the probability for an agent to be placed in one of these areas is very low and it does not influence the fitness significantly.

transformation. In this figure, all curves are located in a quite small area.

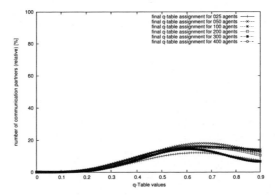

Fig. 7.10. The relative values in the q-tables. All of them are in a similar area, the similarities between the curves are striking.

According to this, the maximum number of entries in the q-table is, within $\approx 15\%$ of the overall number of agents, in the range from 0.6 to 0.7.

We determine an approximation equation ζ to compute the optimal number of relative communication partners. For each q-value, this equation provides the percentage of all agents that an agent should communicate with.

$$\zeta(x) = \begin{cases} 78.569x^5 - 279.24x^4 + 392.8x^3 - 273.3x^2 + 93.9x - 12.5; & x \in [0.4; 1.0] \\ 0; & else \end{cases}$$

The equation is based on the values from many different parametrised runs. The similarity between several functions is shown in figure 7.11.

7.5 Conclusion

In this chapter we presented an evolutionary guided approach to learn qualified communication structures for sets of agents in order to solve an optimisation problem. For a static environment we presented an idea of how the optimal communication structure can be discovered by an algorithm adjusting a communication matrix. For dynamic and random settings we presented a new approach which offers guidelines to create a small set of communication connections. Due to that, only the position in space in relation to some targets is necessary. The optimal or near optimum number of communication partners can be found by our approach. Hence, our approach works exclusively on data that can be locally obtained by the agents. The runtime of this

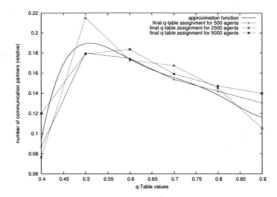

Fig. 7.11. The approximation function ζ can calculate the percentage of agents that should act as communication partners for an arbitrary q-value. This function was derived from multiple simulations. We compared the progressive shape of the function to values *not* used for deriving the function and the similarities are striking. Hence, the function seems to be able to calculate appropriate q-table assignments.

approach depends highly on the overall number of agents. But once a good approximation function has been calculated, the effort for each agent is very low. It only has to store a q-table with a few values.

8

Applications

To make a better mousetrap is *to advise a new kind of mechanism*
whose behaviour is reliable with respect to the high level regularity
'live mouse in, dead mouse out'

J. A. Fodor

In this chapter we try to apply our approaches presented in the previous pages to some 'real world' problems. They came from very different areas, i.e. from the network and the production line domains. Both approaches do not use a one-to-one mapping of the *OPP* to an application but they use the insight into the *OPP* area to find a useful solution.

In the first application, we use the ETS algorithm to enhance access point load in wireless local area networks. We can demonstrate that our approach has no disadvantages in random settings, and in several real world situations it shows very promising behaviour. In the empirical tests, we can triple the overall data rate compared to state of the art approaches.

In the dynamic task allocation we are able to enhance an insect based approach by two local modifications. Both are able to produce significantly better results compared to the state of the art solutions.

8.1 WLan Router Choice

In this section, we present a new approach to enhance the data transfer rates in Wireless Local Area Networks (WLan). Mobile networks become more and more popular in public, especially in companies and in a lot of public buildings. The scope of current research concentrates on the improvement of the transfer speed and the range of coverage, but we will concentrate on a different aspect. In current basic systems, each client chooses the access point with the highest signal strength[1]. This strength is decreasing with the distance. But if the clients are not equally distributed in space according to the access points, such behaviour could lead to an unbalanced workload of the routers which on its part can lead to non-optimal data transfer rates. Our approach adopts some local interactions autonomously done by the clients. Based on these interactions, the clients choose a suitable access point within their communication distance. These interactions or decisions are based on some of the before presented *OPP* algorithms.

8.1.1 WLan Basics

The WLan protocol is a technique to transfer data by radio communication without any cables. The only infrastructure is a set of (mostly) statically installed access points[2] that permit an inter-client communication or access to for example the internet. The access points act as gateways or switches.

The IEEE 802.11b is a standard that describes the functionality of the WLan. In Europe, it operates from 2400.0 MHz to 2483.5 MHz. The bandwidth of one channel is 22MHz and the speed of data transfer is (in this standard) either 1 MBit/s, 2 MBit/s, 5.5 MBit/s or 11 MBit/s [Sik01]. This value depends on the Euclidean distance and the constitution of the environment between two communication partners.

The signal strength P_{rec} for a receiver at a given distance r to the sender with signal strength P_{sig} can be calculated by the equation:

$$P_{rec} = \frac{P_{sig}}{4 \cdot \pi \cdot r^2}$$

This assumes an idealised sender with isotropic[3] behaviour.

8.1.2 Data Transfer Rate

The Shannon-Hartley Theorem describes the maximum possible efficiency of error-correcting methods versus levels of noise interference and data corruption. Hence, the maximum data rate C in dependency on bandwidth B and signal to noise ratio $\frac{S}{N}$ is:

$$C = B \cdot log_2 \left(1 + \frac{S}{N}\right)$$

Due to this Theorem and some parameter values picked up from [Sik01], we obtain a theoretical maximum value of

---

[1] There exist some modifications for connections with very limited data rates. In such situations, clients can decide to chose another reachable access point. This is similar to our approach, but the reconfiguration is non-deterministic and is not based on locality information we are using.

[2] In this section, we do not consider Ad-Hoc networks that do not necessarily need these access points.

[3] An isotropic sender sends the signal in all directions with the same strength.

$$C = 22000000 \cdot log_2(1 + 10^{\frac{8}{10}} Bit/s) \approx 60MBit/s$$

which is far away from the real-world value of 11 MBit/s. Therefore, we developed a function ξ based on manufacturers' instructions[4].

$$\xi(x) = \begin{cases} 11.0 & x \leq 160m; \\ -10.1574\frac{x^3}{550^3} + 44.4147\frac{x^2}{550^2} - 57.4662\frac{x}{550} + 24.2088 & x \in]160, 550]; \\ 0.0 & else \end{cases}$$

With such a function we are able to calculate the data rate for arbitrary distances. The four fixed points picked up from the producer have been converted into one continuous function. We use this function for the calculation of data rates that dependent on the distance to a selected access point.

8.1.3 Assigning *OPP* Algorithms to Access Point Selection

Following the presentation of some basics in the area of WLan networks, an idea will be presented how *OPP* algorithms can enhance the process of access point selection. Suppose, we have a situation as shown in figure 8.1 (left). The clients (filled circles)

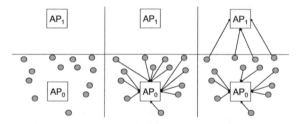

Fig. 8.1. A very bad situation for current WLan settings (left). The dotted line is equidistant to both access points.

are located close to one access point (AP_0) but with not much more distance to a second one (AP_1). Nevertheless, they choose all the same access point since its signal strength is the strongest one (middle). If only some of the clients would choose a different one[5], the data transfer rate may be augmented (right).

We rate our solution by the overall amount of data that can be transferred over all access points. Therefore, we calculate the data that can be transferred from a client to a chosen access point according to its distance to it (see section 8.1.2). This is set in relation to the maximum transfer rate M of the access point. If $C_i = \{c_1, ..., c_{|C_i|}\}$ is the set of clients that have chosen access point AP_i, the data rate for this access point can be calculated by

---

[4] We used a datasheet for the ORiNOCO 11b Client PC Silver network card (see ftp://ftp2.li-life.net/orinoco03/docs/product_data_sheets/client_pccard11b.pdf) for 1 MBit/s (550m), 2 MBit/2 (400m), 5.5 MBit/s (270m) and 11.0 MBit/s (160m).

[5] which is slightly farther away located

$$\mathcal{D}_{AP_i} = \min\left\{M, \sum_{j=0}^{|\mathcal{C}_i|}\xi(\delta(c_j, AP_i))\right\} \tag{8.1}$$

This is a minor modification of the original *OPP* since we do not longer want a uniform distribution of clients onto access points, but our algorithms can cover these modified way of looking at a problem. If clients interact, they have to decide to change their personally chosen access point (probably the nearest one). There are two reasons for doing that. On the one hand, one can change the access point to increase its personal data rate. The additional distance can have less influence on the data rate than the limitations through a large number of clients on the closest one can have. The other reason is a more altruistic one. A client can decide to exchange its access point with another one to increase the sum of the overall data rates, even if this results in a decrease of its own rate. Such an example is shown in figure 8.2.

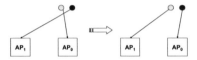

Fig. 8.2. A situation where the overall data rate can be improved. Though the grey client has - after exchanging the access points - a worse rate than before, the increase for the black one is much higher in the new situation. Hence, the maximum overall data rate is higher after the target exchange.

8.1.4 Results

We applied our *OPP* algorithms to different scenarios; their behaviour has been tested on random scenarios and on some artificial scenarios that might occur often in real life, for example in public buildings or during big events. The parameters we have chosen for our simulations are inspired by parameters derived from existing hardware. We compared the overall data rate in different kinds of scenarios. The data rate is determined as in formula 8.1 for each access point individually. The sum of these values should be maximised by the different algorithms. The NTS strategy as shown in section 4.1.3 is the same as the State-of-the-Art algorithms in current WLan solutions. Hence, we try to enhance the data rate with our algorithms. In the following experiments, we applied several basic *OPP* algorithms to the WLan domain. For a better readability, in the following figures only the ETS algorithm - which was the most successful one - is shown as a reference value.

Random Scenarios

We start with the comparison of the different algorithms in random settings. For that purpose, we distribute a large number of clients and several access points randomly in a two-dimensional simulation space and compare the data rates that can be obtained with the different approaches. The results are presented in figure 8.3. The overall data rate used by all clients is summed up. Two cases have been compared. On the one hand the results for high charging clients (each one uses up to 11 MBit/s) and for normal charging clients (up to 1 MBit/s, for example spent for average surfing behaviour). We tried to choose the simulation parameters in a realistic way. The whole

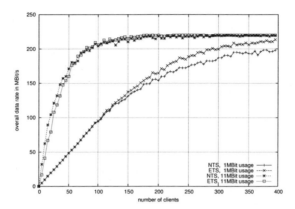

Fig. 8.3. The data rate that can be obtained with the different algorithms. The NTS strategy represents the results by using the State-of-the-Art algorithm. The other approaches use OPP strategies to enhance the result. For these results, we used settings with up to 400 clients and 20 access points. The quality functions show the average data rate from 20 runs.

simulation space is 2000 times 2000 meters large, the access point- and client distance is the maximum value for open space (550 meters, see section 8.1.2). The maximum data rate of the access points is 11 MBit/s.

Limiting Client Communication Radius

The view radius of the clients to exchange information with other clients (and therefore, to enhance the access point choice) was in the previous setting 550 meters, i.e. the same as the WLan communication distance. Hence, the ETS algorithm - for example - can use WLan communication among clients to enhance the result. Another approach would be the usage of another communication protocol for this aim, for example the Bluetooth protocol. Then, the small distance Bluetooth layer can be used to enhance the access point choice and will not strain the WLan. The results of these settings are depicted in figure 8.4. In this scenario, the use of the ETS algorithm shows slightly better results than the NTS algorithm for normal data rates, especially for a high number of clients.

In most of the approaches, ETS was superior to the NTS algorithm. But NTS is the current realisation of access point choice in WLan protocols. Hence, the results can

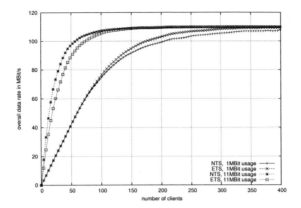

Fig. 8.4. Overall data rate in random scenarios with 10 access points and up to 400 clients using Bluetooth as inter-client communication protocol. The quality function shows the average data rate from 20 runs. An overall data rate of 110 MBit/s ($10 \cdot 11\,MBit\,s^{-1}$) is the maximum value. Then, all 10 access points have a load of 100%. In this setting, no significant difference could be observed.

be improved if our algorithms are applied to this application area. Even if we limit the client communication distance by using a different protocol, the results are competitive. But the improvement is not very strong and does not necessarily justify the additional effort. Next, we examine the performance in some scenarios which are different from the random ones.

Variable Access Point Position

In this section, the influence of an access point's position is examined. Therefore, we create a setting with one fixed access point $AP1$ and a fixed area where all clients are randomly distributed in. This setting should show the influence of access point locations. Then we compare the development of the overall used data rate in the setting (see figure 8.5). The distance between $AP1$ and $AP2$ will be - starting from the maximum distance where some clients are physically able to connect to $AP2$ - stepwise decreased until it is zero. Figure 8.6 points out the altruistic characteristic of the ETS algorithm. Since some clients choose the access point which is farther away, these clients forfeit some amount of their data rate. At the same time, the gain for other clients is much bigger. Thus, the whole group of clients benefits from the altruistic behaviour of some single ones. As we expected, there are large differences in the overall data rate. When $AP2$ enters the scenario, some clients choose it as an access point and hence the overall data rate with ETS is higher than the one with NTS. Not until $AP2$ is closer to some clients than $AP1$, the summed data rate can be doubled with ETS. When $AP2$ is close to or inside the grey area, the results are similar to the random setting (see section 8.1.4). They are presented in figure 8.7.

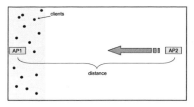

Fig. 8.5. Variable access point scenario description. The overall data rate of the clients in the grey area will be summed up for different distances between $AP1$ and $AP2$.

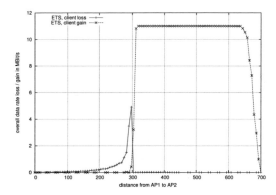

Fig. 8.6. This figure shows the proportion of data rate the clients lose and gain. Only when the access point starts to come close to the client area, some clients forfeit parts of their data rate, while the gain for the whole group is much bigger.

Exhibition environment (Real Life Scenarios)

In the random scenarios' section we could show that some of our approaches are quite similar to the existing solution for access point selection, some results are significantly better. Now we want to examine the quality of our solution in more 'realistic' situations. We start with a short description what we understand by 'realistic' situations.

In most buildings, the position of access points cannot be chosen arbitrarily. It depends on different restrictions, for example architecture, power supply, and cable network connections etc. In such buildings, the distribution of clients depends on other attributes, for example time of the day, usage of rooms, distribution of chairs / desks etc. There exist two possibilities to distribute the access points in a useful way. On the one hand, the network administrator can place as many (or as powerful) access points as possible to cover the worst case client distribution. Or, on the other hand, the access points can be distributed for average case client distributions. Then the data rate for the users has to be limited if there are too many. We concentrate on such cases and analyse the advantage of other access point selection algorithms.

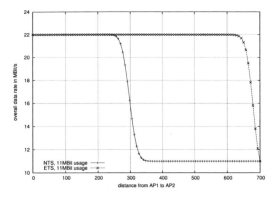

Fig. 8.7. The summed data rate of all clients in the variable access point scenario

This example is inspired by the Hannover fair exhibition. The hall 13 has dimensions and access point distributions similar to the simulation setting presented in figure 8.8. There is no access to other access points located outside of the building. Since the

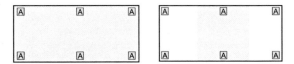

Fig. 8.8. The access point distribution in a building on the Hannover fair ground. Visualised is hall 13. Its dimensions are approx. $220m$ x $110m$. The grey area is the area where the clients are randomly distributed. The A describe where the access points are located.

hall is a closed building, we adjust the maximum WLan data rate to 115 meters.

In two experiments with - besides this data rate - identical parameters as in the random setting we examined the influence of different random distributions on the *OPP* algorithms. The results are visualised in figures 8.9 and 8.10. In these settings the advantage of using *OPP* algorithms for access point choice is obvious. When the clients are distributed uniformly within the room, for both examined data rates no significant difference can be recognised in figure 8.9. This is a validation of the results we obtained in section 8.1.4 for random settings. But when the clients are located in only a part of the whole area, the results in figure 8.10 are more promising. With the NTS algorithm, only two access points will be used whose maximum data rate is exploited by only a few clients. Other strategies, for example the ETS, will use all reachable access points and can hence triple the overall data rate. Such a setting is not as artificial as it might appear, if the hall is subdivided into several areas due to parallel events, such a situation can arise very fast and very often.

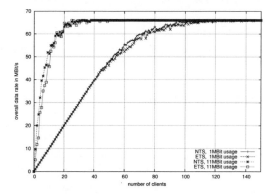

Fig. 8.9. Data rate in a fair building with *arbitrary* distributed clients. The client number is examined from 1 to 150 and the data rates are average values over 20 runs.

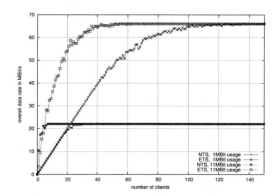

Fig. 8.10. Data rate in a fair building with clients distributed in only a part of the building.

8.1.5 Conclusion

We presented an area where the *OPP* algorithms can easily be applied to a real world problem, the access point selection for groups of clients. Based on local information exchange between the clients which could be realised for example by the Bluetooth interface, the overall data rate could be improved in several scenarios that might occur frequently in reality. We found no scenario where the ETS algorithm was significantly worse than the existing solution.

Of course, one objection to this approach might be that we do not necessarily need this interaction step between the clients. The access point allocation might be done by the access points themselves with orders that will be sent to the clients. But with our approach, we have a highly scalable and dynamic system that works with every router which is currently installed. The more clients make use of it, the better are the

results.
The examinations of the communication amount of *OPP* algorithms showed before
that there is not a great many of messages or interactions necessary to obtain good
results.

8.2 Local Dynamic Task Allocation

Dynamic Task Allocation (DTA) is a common and highly relevant problem for a
variety of challenges that appear in practice. The main idea is to assign tasks to machines. Thereby, the tasks have different types, but these types are initially unknown.
At the same time, the DTA is a decentralised problem. There is no global optimiser or
scheduler that can calculate an optimal solution. In fact, the global problem is already
NP-hard [ZS06].
In this section, we present existing algorithms that cope with this problem field. Then
we modify a successful insect based approach and apply one of our *OPP* algorithms
to it. A machine is represented by a single agent. If the handling time per agent for the
same task is different, we call it a heterogeneous Dynamic Task Allocation problem.
If not, it is a homogeneous one.
Most of the approaches in this section deal with the Painting Facility Problem. This
is a special version of DTA and NP-complete, too. A set of trucks (tasks) have to be
assigned to a set of machines (agents). This assignment should minimise the processing time, i.e. the overall number of machine setups. The task is to paint each truck,
which each have a unknown time of arrival, in a predefined colour. Each machine has
a fixed waiting queue and is capable to paint in any colour, but the change of colour
requires setup time.

8.2.1 Market Based Approaches

There exist several marked based approaches trying to find good solutions for a DTA
instance. The idea behind is the bidding of agents for the different tasks. Mostly, the
single agents specialise on one or several tasks. During runtime, each agent can bid
for a task that might be appropriate for it. The height of the bid depends on the length
of the waiting queue, the priority value of the task, and the necessity of setups.
Let w_j be the priority of truck j, $c_{i,j}$ a decision variable that is '1' if a setup is necessary, '0' otherwise. Δt is the time that will elapse until the task can be processed and
P, C, and L are weights for the components. Then, a bid of agent k for task j can be
expressed by

$$B_k(j) = \frac{P \cdot w_j \cdot (1 + C \cdot c_{i,j})}{\Delta T^L} \tag{8.2}$$

There are some more rules that organise the distribution of the task to agents when
there are bids in the same height. Details about this approach that has been implemented for General Motors and could save in the first nine months approximately one
million dollar can be found in [Mor96] and in [CBTD01].

8.2.2 Insect Based Approaches

In insect colonies, especially in different species of ants, there are different occurrences of task allocation. The reason for agents to specialise on a single task is obvious. The repeated execution of the same action leads to routine with the effects of shorter handling times and quality improvement. In nature, we can distinguish between three types of task allocation [BDT99, Eng05]:

1. *Temporal Polyethism:* The allocation to a special type of tasks depends on the age of the individuals, hence the age of insects working on the same task is more or less the same.
2. *Worker Polyethism:* Differences in the physical abilities of the individuals lead to different processing preferences for each of the existing tasks. In ant colonies, larger or more powerful individuals prefer to for example defend the whole colony.
3. *Individual Variability:* If none of the mentioned categories (neither age nor physical characteristics) are criteria for the task selection, the individuals cultivate their individual preferences due to some feedback or genetically defined processes.

Wilson gives in [Wil84] an example of how such a specialisation could work in colonies of ants (species *Pheidole*): In intact colonies, there are two castes of ants, the Majors and the Minors. The first ones are responsible for the defence of the colony and food storage jobs. The Minors are mainly responsible for brood care. In experiments, Wilson changed the ratio from Minors to Majors from 20:1 to 1:1. The colony was able to compensate the loss of Minors by attaching Majors. These changed their behaviour and allocated tasks previously handled only by Minors. Wilson concluded from these experiments that the abilities for any task in the colony are existent in every ant, independent from the fact whether it is a Major or Minor. Due to the higher size and (as a direct result) the higher energy consumption, the distribution in Major and Minor castes seems to be evolutionary useful.

8.2.3 Mapping to Computer Science

In 1998, Bonabeau, Deneubourg, and Theraulaz [TBD98, BTD98a] used the biological model of Wilson to develop algorithms for the distribution of jobs onto machines. There, each agent (ant) has a fixed threshold level for each possible task. Based on results from Robinson et al. [RPH94], Theraulaz et al. enhanced this system by introducing a dynamic threshold model. Both models have been used in different publications to cope with the painting facility problem. We start with the introduction of these thresholds and will then proceed to the single algorithms that are largely designed on top of each other. The classification of these algorithm is visualised in figure 8.11.

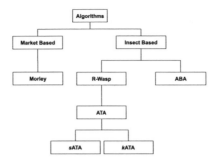

Fig. 8.11. Classification of algorithms dealing with the Painting Facility DTA problem. sATA and kATA are our new approaches presented in the following text.

Fixed Response Thresholds

The idea to control the individual agents' behaviour by fixed thresholds is very simple. The simplest way to explain it is by addressing again the two caste ant colony from Wilson. When we only consider the two tasks *Defence* and *Brood Caring*, Majors and Minors have a different threshold for both of them. The Defence threshold is for Majors very low and for Minors very high, the Brood Caring threshold behaves contrary. Hence, when there is a task with a high stimulus t for the Defence task, all individuals with a threshold $\theta < t$ will take care of this task. These are mainly Majors. If there are not enough ants activated for the task, the stimulus will become stronger and more ants will be activated. If the task is fulfilled, the stimulus will stop.

These thresholds are - as many processes in insect based societies - very fuzzy, i.e. the decisions for tasks are made with different probabilities. In the presented example the tasks will not necessarily be executed if the stimulus is higher than the threshold, but the probability is very high. The same holds for task-stimuli that are below a threshold. Nevertheless, with a small probability the task will be allocated.

The fixed response thresholds model can be formalised in the following way: Suppose, each agent i has an individual threshold $\theta_{i,j}$ for any possible task j, $\theta_i = \{\theta_{i,1}, \theta_{i,2}, ..., \theta_{i,j}\}$. The stimulus of task j is denoted by s_j. There exist several functions that define the probability to choose task j, two often used ones are

$$T_{\theta_i}(s_j) = 1 - e^{-\frac{s_j}{\theta_{i,j}}} \tag{8.3}$$

and

$$T_{\theta_i}(s_j) = \frac{s_j^m}{s_j^m + \theta_{i,j}^m} \tag{8.4}$$

Equation 8.3 is inspired by other behaviours (functions) that appear in nature, but the results of these both formulas are very similar. Bonabeau et al. use equation 8.4 in their experiments, and so do we. An example of the behaviour of the function is shown in appendix B.2.

Dynamic Response Thresholds

In 1994, Robinson et al. found out that the actions of several insect colonies, for example colonies of honey bees, depend on former experiences [RPH94]. This cannot be covered by the fixed response thresholds model. Hence, Theraulaz et al. enhanced this model in [TBD98] by some dynamic aspects. These correct problems of the static model which could not cover several characteristics of the natural system. The four most important ones can be found in [Kem06].

The main idea of the dynamic model is to reduce the threshold for later occurrences of task j if the agent has chosen task j. If $x_{i,j}$ is the percentage fraction of time Δt which agent i spends on processing task j, the new threshold value after Δt time units is

$$\theta_{i,j} = \theta_{i,j} - x_{i,j}\xi\Delta t + (1 - x_{i,j})\rho\Delta t \tag{8.5}$$

In this equation, ξ represents the learning rate and ρ the oblivion rate.

The combination of equation 8.5 and equation 8.4 gives us a probability function that adjusts itself over time and allows the agents to specialise on different task, independent from any preliminary fixing.

Ant-Based Algorithm (ABA)

The Ant-Based Algorithm (ABA) is very similar to the idea of Morley presented before. Again, each agent gives a bid for a task and the task will be assigned to the agent which made the highest one.

$$P_i(j) = \frac{D_{c_j}^2}{D_{c_j}^2 + \alpha \cdot \theta_{i,c_j}^2 + \Delta T^{2\beta}}$$

ΔT is defined as in equation 8.2 and α, β are weights for the components. The global demand D_{c_j} is established for each colour c_j given by the sum of the priorities of the unassigned trucks in each particular colour. Each agent i has a threshold θ for each colour c_j. Regarding the assigned task, the threshold will be updated for all agents. It will be decreased for the agent the task has finally been assigned to

$$\theta_{i,c_j} = \theta_{i,c_j} - \xi; a_i \in \mathcal{A}$$

and increased for all other agents

$$\theta_{k,c_j} = \theta_{k,c_j} + \rho \quad \forall a_k \in \mathcal{A}, k \neq i$$

This approach showed slightly better results compared to the marked based approach we presented in section 8.2.1. All details of this approach can be found in [CBTD01].

R-Wasp

The R-Wasp approach is a bit different from the ABA. It is inspired by the competition behaviour of wasps in a wasp colony. There are two main steps. In the first one, each agent w bids for a task j with stimulus S_j corresponding to the dynamic threshold model introduced in equation 8.4

$$P(bid|\theta_{w,j}) = \frac{S_j^2}{S_j^2 + \theta_{w,j}^2}$$

This is the probability for agent w to bid for task j. Similar to ABA, the thresholds will be adjusted after each task assignment, but this update function is only applied to the single agent the task has been assigned to. First, the threshold for the assigned task j will be decreased

$$\theta_{w,j} = \theta_{w,j} - \xi$$

and then the thresholds for all other tasks will be increased

$$\theta_{w,i} = \theta_{w,i} + \rho \quad \forall i \neq j$$

There are some more rules that modify the thresholds if an agent is idle, for a detailed description we refer to [CS04].

In the second step one agent out of all that bid for the task will be selected. This is done by calculating a force value that depends on the necessary times for setups (T_s) and the length of the waiting queue (T_p).

$$F = 1.0 + T_p + T_s$$

The value is increased by '1.0' to avoid a division by zero. Following this local calculation, the agents are in a pairwise competition. The probability to win depends on the force values:

$$P(\text{agent 1 wins}|F_1, F_2) = \frac{F_2^2}{F_1^2 + F_2^2}$$

This is a very strange and complicated selection process that can probably be simplified by a single, roulette wheel like stochastic step, but we implemented it exactly as described in the original paper [CS04] to obtain comparable results.

This algorithm shows on different test problems better results compared to the ABA and the market based approach we presented before.

Ant-Task Allocation Algorithm (ATA)

In all the former approaches, the currently processed job had an influence on the task preferences. Ghizzioli et al. [GNBD05] modified this by considering the tasks in the waiting queue, especially the last one. If this task has the same colour as the new task, no additional setup is needed. This is the most important and relevant modification the author has done in his work; three other ones can be found in [GNBD05].

8.2.4 Adopting *OPP* Ideas

The heuristics presented before for the painting facility problem base all on ideas that have been motivated by insect behaviours. But these behaviours have only been assigned to the single *machines*. We will now examine if the solution quality can be enhanced if we allow the single *tasks* to interact. These interactions are motivated by some algorithms developed for the *OPP*.

The first approach tries to enhance the solution by local modifications of the waiting queues. The second one adds a reservoir which allows to update the machines' decisions due to additional information obtained from other agents.

sorting ATA (*s*ATA)

When considering the most successful of the before presented insect based algorithms, some not optimal waiting queues could be observed. An example for such an unfavourable queue is shown in figure 8.12. By local task exchanges, the number of setups can be easily reduced without losing the scalability of the system. The whole

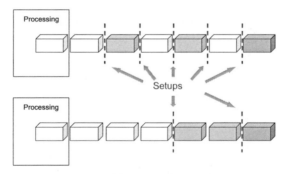

Fig. 8.12. An example for a not optimised waiting queue and its corresponding optimised version.

queue optimisation algorithm is very similar to the ETS presented in section 4.2.3. The task distribution is close to the optimum after initialisation. The setup minimisation objective is similar to the distance minimisation objective. A basic version of this idea can be implemented as shown in Algorithm 13. In this algorithm, both costs for exchanging targets between agent and costs for the stimuli are not considered. One possibility is to embed these costs into the decision in line 6.

k-reservoir ATA (*k*ATA)

In this approach, an additional step following the distribution algorithm is added. When a task has been assigned to one machine, it will not directly be placed in the waiting queue but in a reservoir. There it remains for a predefined time. During this

---

Algorithm 13 Local Queue Optimisation for ATA

1: **procedure** OPTIMISEQUEUESORTING(AgentSet A, NeighbourhoodSize k)
2:    **repeat**
3:        $changed \leftarrow false$
4:        **for all** agents a in AgentSet A (parallel) **do**
5:            $a.examineNeighbourhood(k)$
6:            **if** (numberOfSetups can be reduced by local exchange with agent b) **then**
7:                $a.exchangeQueuePosition(b)$
8:                $changed \leftarrow true$
9:            **end if**
10:       **end for**
11:   **until** (changed $\neq$ true)
12: **end procedure**

---

stopover, it can interact with other agents and change its decision from one machine to another. Thus, each task has the additional abilities to communicate / interact with other tasks in a locally defined neighbourhood.

The whole algorithm is visualised in figure 8.13. The first assignment is done as seen before by the insect based ATA algorithm that generates good results. The tasks will be stored in the reservoir for t time steps. During this stay, they can exchange assignments with other tasks. The objective is to minimise the overall number of setups. In the given example, the reassignment of the two tasks can reduce the number of setups for machines 1 and 2 in each case by 1. Such a situation can occur if a machine is forced to process a task it is not specialised on due to a high task stimulus.

One important question is the definition of the neighbourhood of one task. In the

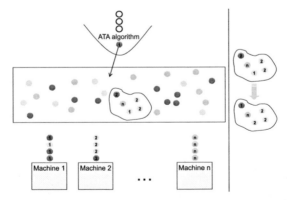

Fig. 8.13. Modus operandi of the kATA algorithm. After the assignment of tasks by the ATA algorithm, each task can update its decision by exchanging targets with neighboured tasks in the reservoir. One example is provided on the right side. The number of setups can be reduced by these local interactions.

painting facility problem, the Euclidean distance between trucks can still define if tasks are neighboured. But in general, several neighbourhoods are possible. If the reservoir is seen as a long intermediate waiting queue Q, a neighbourhood can be defined considering the time a task has been placed in Q.

We implemented only a preliminary version of this algorithm to check if the concept might work. But this version already provides very interesting and often very good results. These are presented in the next section.

8.2.5 Results

The new approaches have been implemented in a bachelor thesis under our supervision [Kem06]. We could reproduce the ATA results from Ghizzioli et al. [GNBD05] and obtained nearly the same results. To compare our approaches with the existing ones, we used the same parameter settings for the experiments. Additionally, we compared the quality of the approaches for varying agent and task numbers. The benchmark results are listed in table 8.1. These values are visualised in figures 8.14 and

machines	ATA MS	ATA S	sATA MS	sATA S	kATA MS	kATA S
6	4169.81	247.64	**3476.65**	**178.07**	4940.62	226.63
10	2231.64	120.34	**1928.32**	89.72	2391.90	**68.09**
11	1962.07	102.38	**1701.83**	76.06	2324.72	**47.47**
12	1738.24	87.34	**1520.62**	65.36	1967.94	**30.85**
18	752.74	13.51	**738.05**	11.59	1062.42	**6.62**
24	535.20	4.30	**533.08**	**3.87**	883.81	8.12
48	463.92	4.43	**460.65**	**3.97**	643.96	9.82

Table 8.1. The results of the sATA and kATA approaches compared to ATA. In literature, the results for settings with 24 agents are listed. In all settings, our approaches could outperform the original algorithm. The shortest values for makespan (MS) and the lowest number of setups (S) is emphasised with bold numbers.

8.15. After 210 time steps, the probability distribution of the processed task types changed. This was introduced by Ghizzioli and adopted in our simulations to test the ability to specialise and to re-specialise for task types. For all examined numbers of machines, the sATA has the minimal makespan. The improvement compared to the standard ATA is up to 16.6 percent. The fewer machines are available, the higher is the difference. The number of setups could likewise be reduced significantly with our modifications.

The kATA approach shows a different behaviour. In several runs, the results are better than the ones obtained by the other approaches, especially regarding the number of setups for small numbers of machines. For high number of machines[6], the number of setups can be more than doubled (for example with 48 machines the increase of setups is approximately 120%). The makespan is always higher in the kATA approach compared to ATA and sATA.

The remaining results for varying task numbers can be found in appendix C.2.

---

[6] these are the easier problems since there are more choices

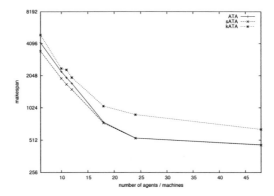

Fig. 8.14. The makespan results for ATA, *s*ATA, and *k*ATA algorithms. 2016 tasks have been distributed on different numbers of machines. For better readability, the y-axis has been set to logarithmic scale. The presented values are averages from 100 runs each.

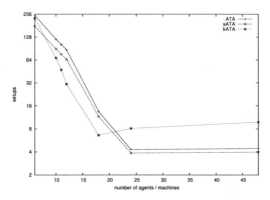

Fig. 8.15. The setup results for ATA, *s*ATA, and *k*ATA algorithm. 2016 tasks have been distributed on different numbers of machines. For better readability, the y-axis has been set to logarithmic scale. The presented values are averages from 100 runs each.

8.2.6 Conclusion

In this second application example we presented an idea how to improve Dynamic Task Allocation heuristics. The basic ideas of these heuristics base on insect behaviours, particular on task specialisation in wasp and ant colonies. Three ideas to solve the Painting Facility Problem that are designed on top of each other have been introduced and the most successful one has been modified. The tasks in the waiting queue have been considered as agents for the *OPP*. By applying the ETS algorithm to these agents, the overall result could be significantly improved. We showed that the improvement is strong for settings with noticeable more task types than machines.

8.3 Fire Fighting with Unmanned Aerial Vehicles (UAV)

In the previous presented applications, the *OPP* algorithms could only be used partly or with modifications to improve state of the art approaches. In this section, a direct application will be presented as a proof of concept.

8.3.1 Motivation

In current robotic and autonomous vehicles research, the area of Unmanned Aerial Vehicles (UAVs) becomes increasingly important. Very good results could be reached especially in the military area where drones are utilised more and more often and with more and more success. We want to put the focus on some civil applications dealing with similar topics. Assuming that there exists an extensive number of UAVs which can carry some amount of water to fight fires in unfriendly terrains, these machines can and should not be controlled by a single instance. One can imagine a broadcast about possible seats of fire and their positions, but the choice which UAV should reach for which seat of fire has to be made very fast and dynamically. In a good system, the decisions will be made by the UAVs themselves, based on only local information to reduce computing and communication power.

8.3.2 Advantages and Mapping of *OPP* Algorithms

The *OPP* objectives fit perfectly into this problem field. The uniform distribution of the UAVs onto the single seats of fire would guarantee an optimal fire fighting on all peril points at the same time. The distance minimisation can guarantee a fast starting of the fire fighting.

There is one difference to the *OPP* settings we considered in the chapters before. The agents (or UAVs) are no longer located on a fixed position and only decide for a target, but they decide for a target and move to it afterwards. If the decision process is done uncoupled from the moving phase in a first step, all algorithms can be used. But the overall result might be different if the decision for a target can be reconsidered during the heading for it. We analysed this enhancement in section 4.4. Hence, all the results presented there could be mapped to this application.

9

Conclusions and Future Work

Great is the art of beginning,
but greater is the art of ending.

Henry Wadsworth Longfellow

In this work, we presented a number of coordination mechanisms for large groups of agents. With these approaches, the introduced partitioning problem could be solved with different qualities. Because of varying agent characteristics, the solutions cover a wide variety of conceivable problem instances. In detail, heuristics have been developed with focus on individual behaviour, group organisation, and inter-agent communication.

As a theoretical background, we analysed in chapter 2 a graph structure, the k-neighbourhood graphs. Some communication bounds could be proven. In an empirical analysis, a sufficient number of communication partners in such graphs has been determined and the advantages of kNGs compared to radial communication have been shown.

The aspect of individual behaviour was the focus in chapter 4 and chapter 5. Both algorithms for non-communicating and local communicating agents have been developed. For all of them, some worst cases have been presented. One of the most promising algorithms, the simple but effective ETS, has been examined in a more detailed way. In this context, reasonable numbers for communication partners and a function that governs the amount of randomness could be determined. With the cellular automata approach, ideas from a different area could be mapped to the partitioning task. After the introduction of some new operators for the installed genetic algorithm, the offline learning performance could be enhanced to a more acceptable level. Though the results are only slightly better than the very basic algorithms presented before, this idea can easily be used for other optimisation problems by an adjustment of the fitness function. The improvement of the set of rules might be subject to future work.

Based on research in the field of economic simulations, the organisational aspects in groups of agents have been focussed upon in chapter 6. Organisation forms that can be observed in economics constitute the basis for inter-agent organisations. Two approaches have been introduced. The first forms organisations among arbitrary located agents and shows very fast convergence behaviour. With the second approach, the concept of holonic agents was incorporated. Based on only local information, the agents are able to form an almost completely connected organisation in which the single organisation heads can obtain an abstract global knowledge of the distribution of the agents and their individually preferred target. In this context, future research could carry on the robustness aspect. That might be important since the failure of an (sub-)organisation head could result in difficulties for the subordinated agents in the concerned (sub-)organisation.

Since most of the successful strategies presented so far make use of communication, chapter 7 focuses on this aspect. Here, we have not designed strategies according to given communication constraints as we did in the other chapters, but we developed a communication structure tailored to the given optimisation problem. Optimal numbers for communication partners have been found with the help of an evolutionary algorithm. The decision for the appropriate learned number is directed by each individual agent's position which can be computed autonomously based only on the particular knowledge. This could lead to an extensive amount of communication connections in single regions which might be unmanageable in real systems. Hence, we additionally examined the behaviour of this approach when some limits or upper bounds are given. The communication is unequally distributed in space which makes the overall performance comparatively good.

The application area is rooted in several questions which appeared during conferences and discussions regarding the adaptability of the *OPP* to real world problems. Two examples have been presented where the application of the *OPP* could significantly enhance approaches from different areas. In the wireless LAN environment, the access point choice could be improved by simple client interactions. Several settings have been presented where this leads to a more uniformly distributed workload of the access points. In the field of dynamic task allocation, a special instance, the painting facility task, could be enhanced by two modifications inspired by the *OPP*. In the first one, intelligent local permutations could outperform the existing approaches. The second modification introduced a reservoir where the tasks could locally enhance the decisions before being assigned to the varnishing machines. Especially the latter should be subject to future research since the solution presented here is not exhaustively optimised. Nevertheless, in several instances it could outperform existing approaches. In the end of this thesis, we introduced a further idea where to use *OPP* algorithms in the area of moving agents.

We presented different heuristics for the *OPP* by all approaches. It is hard to say in general which of the developed algorithm is best since it depends on the given agents' abilities and other application-dependent characteristics. But with these ideas several possible starting points are offered. Nevertheless, the ETS algorithms using the determined parameters might be a good choice to start with.

A

Proofs

A.1 Maximum of Product Function

Theorem
A function $f = x_1 \cdot x_2 \cdot \ldots \cdot x_n$ with $x_1 + x_2 + \ldots + x_n = C$ is maximal for $x_1 = x_2 = \ldots = x_n$, i.e. $x_1 = x_2 = \ldots = x_n = \frac{n}{C}$ with $x_i \neq 0 \ \forall i \in \{1, 2, \ldots, n\}$.

Proof
To proof this theorem, we calculate the gradient $\nabla \cdot f$:

$$\nabla f(x_1, x_2, \ldots, x_n) = \begin{pmatrix} \frac{\partial f}{\partial x_1} \\ \frac{\partial f}{\partial x_2} \\ \vdots \\ \frac{\partial f}{\partial x_n} \end{pmatrix} = \begin{pmatrix} x_2 \cdot x_3 \cdot \ldots \cdot x_n \\ x_1 \cdot x_3 \cdot \ldots \cdot x_n \\ \vdots \\ x_1 \cdot \ldots \cdot x_{n-2} \cdot x_n \\ x_1 \cdot \ldots \cdot x_{n-2} \cdot x_{n-1} \end{pmatrix} = \begin{pmatrix} v_1 \\ v_2 \\ \vdots \\ v_n \end{pmatrix}$$

This gradient is (0) on the maximum of f, therefore it holds for all elements v_i in ∇f that v_i has to be zero.

From $v_1 = 0 = v_2$ we can conclude that $x_1 = x_2$, or, in general, we can transform $v_i = 0 = v_{i+1}$ to $x_i = x_{i+1} \forall i \in \{1, 2, \ldots, (n-1)\}$. It is easy to see that all x_i's have to be equal to fulfil these objectives.

A.2 Properties of q-Intervals

Theorem 14. *All points in space that have one specific q-value according to two targets t_1 and t_2 on positions (t_{1_x}, t_{1_y}) and (t_{2_x}, t_{2_y}) lie on the circles C_1 and C_2 with centre points*

$$M_1 = \left(\left(\frac{t_{2_x} - q^2 \cdot t_{1_x}}{1 - q^2} \right), \left(\frac{t_{2_y} - q^2 \cdot t_{1_y}}{1 - q^2} \right) \right)$$

$$M_2 = \left(\left(\frac{t_{1_x} - q^2 \cdot t_{2_x}}{1 - q^2} \right), \left(\frac{t_{1_y} - q^2 \cdot t_{2_y}}{1 - q^2} \right) \right)$$

and radius

$$r = \frac{q \cdot D}{(1 - q^2)}$$

Proof. q is calculated for an arbitrary point $p = (x, y)$ by the formula

$$q = \frac{dist.\ to\ nearest\ target}{dist.\ to\ farthest\ target}$$

This is the same definition as in equation 7.1, reduced to two targets. Without loss of generality we assume that target t_2 is the nearest one and t_1 the farthest one. Therefore, q can be expressed by:

$$q = \frac{\sqrt{(x - t_{2_x})^2 + (y - t_{2_y})^2}}{\sqrt{(x - t_{1_x})^2 + (y - t_{1_y})^2}}$$

This can be transformed to:

$$\left(\sqrt{\frac{q^2 \cdot t_{1_x}^2 + q^2 \cdot t_{1_y}^2 - t_{2_x}^2 - t_{2_y}^2}{1 - q^2}} + \left(\frac{t_{2_x} - q^2 \cdot t_{1_x}}{1 - q^2} \right)^2 + \left(\frac{t_{2_y} - q^2 \cdot t_{1_y}}{1 - q^2} \right)^2 \right)^2$$

$$= \left(x - \left(\frac{t_{2_x} - q^2 \cdot t_{1_x}}{1 - q^2} \right) \right)^2 + \left(y - \left(\frac{t_{2_y} - q^2 \cdot t_{1_y}}{1 - q^2} \right) \right)^2$$

and this can be simplified to a standard circle equation:

$$\left(\frac{q \cdot D}{(1 - q^2)} \right)^2 = \left(x - \left(\frac{t_{2_x} - q^2 \cdot t_{1_x}}{1 - q^2} \right) \right)^2 + \left(y - \left(\frac{t_{2_y} - q^2 \cdot t_{1_y}}{1 - q^2} \right) \right)^2$$

These are centre point M_1 and radius r in our theorem. If target t_1 is the nearest one, we obtain the other centre point M_2 by using the same transformations.   □

B

Function characteristics

B.1 ψ Function for Mapping

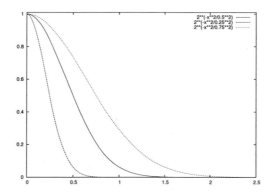

Fig. B.1. Psi functions with different parameters

B.2 Response Probability Function

Here, the function $T_{\theta_i}(s_j) = \frac{s_j^m}{s_j^m + \theta_{i,j}^m}$ with $m = 2$ is shown for different threshold values ($\theta_{i,j} = \{5, 10, 15, 20, 50\}$).

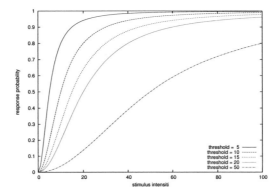

Fig. B.2. Probability to act on a stimulus

C

Additional Simulation Results

C.1 Final Fitness Values for Different Parameters

# agents	q-table entries	final fitness value
25	10	0.81290
50	01	0.78493
50	10	0.83313
50	20	0.83147
100	01	0.79957
100	05	0.84552
100	10	0.84702
200	01	0.81192
200	10	0.85937
300	10	0.86567
400	10	0.87057
1000	10	0.86534
2000	10	0.86857

Table C.1. Final fitness values for different parameters. The values represent the fitness after 3000 generations.

C.2 sATA and kATA Results

In this section, the makespan and number of setups for the three algorithms ATA, sATA, and kATA will be presented. The results are calculated for different numbers t of tasks ($t \in \{126, 1008, 2016, 4032\}$). All values are average values from 100 runs.

C.2.1 sATA and kATA Results for 126 tasks

machines	ATA MS	ATA S	sATA MS	sATA S	kATA MS	kATA S
6	228.29	4.69	**223.91**	**4.20**	311.04	10.52
10	155.20	3.23	**147.13**	**2.74**	265.00	6.64
11	146.39	3.03	**138.17**	**2.64**	247.85	6.10
12	139.95	2.81	**132.60**	**2.48**	246.20	5.47
18	100.42	2.25	**100.39**	**1.95**	215.11	3.92
24	85.77	2.11	**84.44**	**1.77**	156.06	2.93
48	86.14	1.63	**82.43**	1.43	101.12	**1.29**

Table C.2. Results of the sATA and kATA approaches compared to ATA (126 tasks)

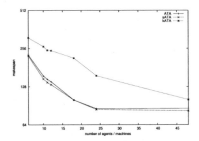

Fig. C.1. Makespan results for ATA, sATA, and kATA algorithm (126 tasks)

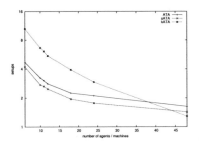

Fig. C.2. Setup results for ATA, sATA, and kATA algorithm (126 tasks)

C.2.2 sATA and kATA Results for 1008 tasks

machines	ATA MS	ATA S	sATA MS	sATA S	kATA MS	kATA S
6	4169.81	247.64	**3476.65**	**178.07**	4940.62	226.63
10	2231.64	120.34	**1928.32**	89.72	2391.90	**68.09**
11	1962.07	102.38	**1701.83**	76.06	2324.72	**47.47**
12	1738.24	87.34	**1520.62**	65.36	1967.94	**30.85**
18	752.74	13.51	**738.05**	11.59	1062.42	**6.62**
24	535.20	4.30	**533.08**	**3.87**	883.81	8.12
48	463.92	4.43	**460.65**	**3.97**	643.96	9.82

Table C.3. Results of the sATA and kATA approaches compared to ATA (1008 tasks)

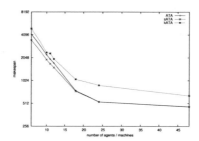

Fig. C.3. Makespan results for ATA, sATA, and kATA algorithm (1008 tasks)

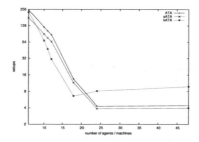

Fig. C.4. Setup results for ATA, sATA, and kATA algorithm (1008 tasks)

C.2.3 sATA and kATA Results for 4032 tasks

machines	ATA MS	ATA S	sATA MS	sATA S	kATA MS	kATA S
6	8819.55	544.71	**7194.49**	**381.98**	10733.63	503.28
10	5033.35	300.08	**4171.40**	213.59	6493.10	**182.02**
11	4499.84	264.93	**3739.91**	188.67	5423.72	**117.28**
12	4067.10	236.86	**3377.92**	167.60	4322.73	**76.35**
18	2130.86	97.94	**1907.37**	75.08	2231.66	**10.77**
24	990.04	7.63	**980.49**	**6.44**	1916.76	13.64
48	868.24	7.03	**863.57**	**6.26**	1000.36	16.79

Table C.4. Results of the sATA and kATA approaches compared to ATA (4032 tasks)

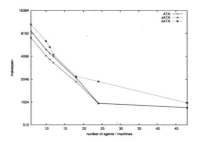

Fig. C.5. Makespan results for ATA, sATA, and kATA algorithm (4032 tasks)

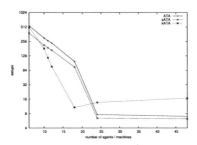

Fig. C.6. Setup results for ATA, sATA, and kATA algorithm (4032 tasks)

References

[ABK96]   D. Andre, F. H. Bennett, and J. Koza. Discovery by genetic programming of a cellular automata rule that is better than any known rule for the majority classification problem. In *Proceedings of the First Annual Conference on Genetic Programming*. MIT Press, 1996.

[AM76]   W. Ashcroft and N.D. Mermin. *Solid State Physics*. W.B. Saunders, Philadelphia, 1976.

[BBP98]   F. Bagnoli, N. Boccara, and P. Palmerini. Phase transitions in a probabilistic cellular automaton with two absorbing states. In *Phys. Rev. E 71*, Singapore: World Scientific, 1998. Lecture notes of the Summer School on Biotechnology held at Torino (Italy) in June 1996.

[BDMA02]   David J. Bruemmer, Donald D. Dudenhoeffer, Mark D. McKay, and Matthew O. Anderson. A robotic swarm for spill finding and perimeter formation. In *Spectrum 2002*, Reno, NV, 2002.

[BDT99]   Eric Bonabeau, Marco Dorigo, and Guy Theraulaz. *Swarm Intelligence - From natural to artificial Systems*. Oxford University Press, 1999.

[Bec05]   Lars Beckmann. Evolutionäre Entwicklung und Optimierung von Kommunikationsstrukturen zur Koordination von Agenten. Master thesis, University of Paderborn, 2005.

[Ben04]   Gerardo Beni. From swarm intelligence to swarm robotics. In *Swarm Robotics WS 2004*, pages 1–9, Berlin Heidelberg, 2004. Springer, LNCS.

[BF98]   N. Boccara and H. Fuks. Cellular automaton rules conserving the number of active sites. In *Journal of Physics A: Math. Gen. 31*, pages 6007–6018, 1998.

[BF02]   N. Boccara and H. Fuks. Number-conserving cellular automaton rules. In *Fundamenta Informaticae 52*, 2002.

[BH00]   Tucker Balch and Maria Hybinette. Social potentials for scalable multirobot formations. In *Proceedings of the IEEE International Conference on Robotics and Automation (ICRA-2000)*, San Francisco, 2000.

[Bis03]   B. E. Bishop. On the application of redundant manipulator techniques to the control of platoons of autonomous vehicles. *IEEE Transactions on Systems, Man and Cybernetics Part A, Special Issue on Collective Intelligence*, 33(5):608–615, 2003.

[BK02]   C. Belta and V. Kumar. Trajectory design for formations of robots by kinetic energy shaping. In *Proceedings of the IEEE International Conference on Robotics and Automation (ICRA'02)*. IEEE Press, 2002.

[Bla99]   Susan Blackmore. *The Meme Machine*. Oxford University Press, 1999.

[Bor86]   Jay Boris. A vectorized 'near neighbors' algorithm of order n using a monotonic logical grid. *J. Comput. Phys.*, 66(1):1–20, 1986.

[BPJ02]   A. Byde, C. Preist, and N. Jennings. Decision procedures for multiple auctions. In *Proceedings of the 1st Joint International Conference on Autonomous Agents and Multi-Agent Systems*, 2002.

[BSS03]   Erkin Baheci, Onur Soysal, and Erol Sahin. A review: Pattern formation and adaption in multi-robot systems. Technical Report CMU-RI-TR-03-43, Robotics Institute, Carnegie Mellon University, Pittsburgh, Pennsylvania 15213, 2003.

[BTD$^+$97]   Eric Bonabeau, Guy Theraulza, Jean-Louis Deneubourg, Serge Aron, and Scott Camazine. Self-organization in social insects. Working Papers 97-04-032, Santa Fe Institute, 1997. available at http://ideas.repec.org/p/wop/safiwp/97-04-032.html.

[BTD98a]   E. Bonabeau, G. Theraulaz, and J. Deneubourg. Fixed response thresholds and the regulation of division of labour in insect societies. *Bulletin of Mathematical Biology*, pages 753–807, 1998.

[BTD+98b] E. Bonabeau, G. Theraulaz, J.-L. Deneubourg, N. R. Franks, O. Rafelsberger, J.-L. Joly, and
S. Blanco. A model for the emergence of pillars, walls and royal chambers in termite nests. *Philosophical Transactions of the Royal Society B: Biological Sciences*, 353(1375):1561–1576, 1998.

[BW93]   G. Beni and J. Wang. Swarm intelligence in cellular robotic systems. In *Robots and Biological Systems: Towards a new Bionics?*, pages 703–712. Springer, 1993.

[Cal93]   P.B. Callahan. Optimal parallel all-nearest-neighbours using the well-separated pair decomposition. In *Proceedings of the 34th IEEE Symposium Foundations of Computer Science*, pages 332–340, 1993.

[CBTD01]   M. Campos, E. Bonabeau, G. Theraulaz, and J.-L. Deneubourg. Dynamic scheduling and division of labor in social insects. *Adaptive Behavior*, 8(2):83–92, 2001.

[CDF+01]   S. Camazine, J.-L. Deneubourg, N.R. Franks, J. Sneyd, G. Theraulaz, and E. Bonabeau. *Self Organization in Biological Systems*. Princeton University Press, NJ, 2001.

[CDM92]   A. Colorni, M. Dorigo, and V. Maniezzo. Distributed optimization by ant colonies. In *Proceedings of the 1st European Conference on Artificial Life*, pages 134–142, 1992.

[CH97]   Alper K. Caglayan and Colin G. Harrison. *Agent Sourcebook*. John Wiley & Sons, Inc, Cambridge, 1997.

[CS04]   Vincent A. Cicirello and Stephen F. Smith. Wasp-like agents for distributed factory coordination. *Autonomous Agents and Multi-Agent Systems*, 2004.

[CS06]   Francesc Comellas and Emili Sapena. A multiagent algorithm for graph partitioning. In *EvoWorkshops*, pages 279–285, 2006.

[CSY99]   H.F. Chau, L.W. Siu, and K.K. Yan. One dimensional n-ary density classification using two cellular automaton rules. In *International Journal of Modern Physics C, 10(5)*, pages 883–889, 1999.

[Dar59]   Charles Darwin. *The Origin of Species by Means of Natural Selection*. John Murray, London, 1859.

[Daw89]   Richard Dawkins. *The Selfish Gene*. Oxford University Press, 2nd edition, 1989.

[Dec00]   Ethan H. Decker. Landscape ecology and macroscopic dynamics: Self-organizing systems. In *Biology 576*, 2000.

[DG97]   M. Dorigo and L.M. Gambardella. Ant colony system: A cooperative learning approach to the traveling salesman problem. *IEEE Transactions on Evolutionary Computation*, 1(1):53–66, 1997.

[Dic06]   Merriam Webster Online Dictionary. Agent, 2006. [Online; accessed 26-June-2006].

[DJP+94]   Elias Dahlhaus, David S. Johnson, Christos H. Papadimitriou, P. D. Seymour, and Mihalis Yannakakis. The complexity of multiterminal cuts. *SIAM J. Comput.*, 23(4):864–894, 1994.

[DK02]   Xavier Defago and Akihiko Konagaya. Circle formation for oblivious anonymous mobile robots with no common sense of orientation. In *Proceedings of the POMC'02*, Toulouse, France, 2002. ACM.

[EA96]   Joshua M. Epstein and Robert L. Axtell. Growing artificial societies: Social science from the bottom up. *Complex Adaptive Systems*, 1996.

[EH01]   M. Egerstedt and X. Hu. Formation constraint multi-agent control. *IEEE Transactions on Robotics and Automation*, 17(6):947–951, 2001.

[Eng05]   Andries P. Engelbrecht. *Fundamentals of Computational Swarm Intelligence*. John Wiley and Sons Ltd., England, 2005.

[Epp96]   David Eppstein. Spanning trees and spanners. Technical Report 96-16, University of California, Dept. Information and Computer Science, 1996.

[EPY97]   David Eppstein, Michael S. Paterson, and Frances F. Yao. On nearest-neighbor graphs. *GEOMETRY: Discrete & Computational Geometry*, 17, 1997.

[ES03]   A.E. Eiben and J.E. Smith. *Introduction to Evolutionary Computing*. Springer, 2003.

[Eym00]   Torsten Eymann. *AVALANCHE - Ein agentenbasierter dezentraler Koordinationsmechanismus für elektronische Märkte*. Phd thesis, University of Freiburg, Germany, 2000.

[Fer99]   Jacques Ferber. *Multi-agent systems: An introduction to distributed artificial intelligence*. Addison-Wesley, Harlow, England [u.a.], 1999.

[Fie04]   Jonathan E. Fieldsend. Multi-objective particle swarm optimisation methods. Technical Report 419, Department of Computer Science, University of Exeter, 2004.

[FM02]   Jakob Fredslund and Maja J. Matarić. A general algorithm for robot formation using local sensing and minimal communication. *IEEE Transactions on Robotics and Automation, special issue on Advances in Multi-Robot Systems*, 18(5):837–846, 2002.

[FMSY02]   Kenichi Fujibayashi, Satoshi Murata, Ken Sugawara, and Masayuki Yamamura. Self-organizing formation algorithm for active elements. In *Proceedings of the 21st IEEE Symposium on Reliable Distributed Systems (SRDS'02)*, page 416, Washington, DC, USA, 2002. IEEE Computer Society.

[FPSW99]   Paola Flocchini, Giuseppe Prencipe, Nicola Santoro, and Peter Widmayer.   Hard tasks for weak robots: The role of common knowledge in pattern formation by autonomous mobile robots. In *Proceedings of the 10th International Symposium on Algorithms and Computation(ISAAC '99)*, pages 93–102, London, UK, 1999. Springer-Verlag.

[FPSW01]   Paola Flocchini, Giuseppe Prencipe, Nicola Santoro, and Peter Widmayer.   Gathering of asynchronous oblivious robots with limited visibility. In *Proceedings of the 18th International Symposium on Theoretical Aspects of Computer Science (STACS 2001)*, pages 247–258, Dresden, Germany, 2001. Springer-Verlag Berlin, Heidelberg.

[GBPW05a]   Andreas Goebels, Hans Kleine Büning, Steffen Priesterjahn, and Alexander Weimer. Multi target partitioning of sets based on local information. In *Proceedings of the fourth IEEE Workshop on Soft Computing as Transdisciplinary Science and Technology (WSTST'05)*, pages 1309–1318, Muroran, Japan, 2005. Springer, Germany.

[GBPW05b]   Andreas Goebels, Hans Kleine Büning, Steffen Priesterjahn, and Alexander Weimer. Towards on-line partitioning of agent sets based on local information. In *Proceedings of the International Conference on Parallel and Distributed Computing and Networks (PDCN'05)*, pages 674–679, Innsbruck, Austria, 2005.

[GGH+01]   J. Gao, L. Guibas, J. Hershberger, L. Zhang, and A. Zhu. Geometric spanner for routing in mobile networks. In *Proc. ACM MobiHoc01*, pages 45–55, 2001.

[GJS76]   M. R. Garey, D.S. Johnson, and L. Stockmeyer. Some simplified np-complete graph problems. *Theoretical Computer Science*, 1(3):237–267, 1976.

[GKL78]   P. Gacs, G.L. Kurdyumov, and L. Levin. One-dimensional uniform arrays that wash out finite islands. In *Problemy Peredachi Informatsii 12*, pages 92–98, 1978.

[GNBD05]   R. Ghizzioli, S. Nouyan, M. Birattari, and M. Dorigo. An ant-based algorithm for the dynamic task allocation problem. Technical Report TR/IRIDIA/2005-31, Universite Libre de Bruxelles, 2005.

[Goe03]   Andreas Goebels. Strukturbildung durch koordinierte und evolutionäre Partikelschwärme. Diploma thesis, Universität Paderborn, 2003.

[Goe06a]   Andreas Goebels. Learning useful communication structures for groups of agents. In *Proceedings of the IFIP Conference on Biologically Inspired Cooperative Computing (BICC'06)*, pages 125–135, Santiago de Chile, Chile, 2006. Springer, Germany.

[Goe06b]   Andreas Goebels. A mapping function to use cellular automata for solving mas problems. In *Proceedings of the IEEE International Conference on Natural Computation (ICNC'06)*, pages 53–62, Xi'an, China, 2006. Springer, Germany.

[Goe06c]   Andreas Goebels. Studies on neighbourhood graphs for communication in multi agent systems. In *Proceedings of the IEEE International Conference on Natural Computation (ICNC'06)*, pages 456–465, Xi'an, China, 2006. Springer, Germany.

[Gol89]   D. E. Goldberg. *Genetic Algorithms in Search, Optimization and Machine Learning*. Addison Wesley, 1989.

[GP03]   Vincenzo Gervasi and Giuseppe Prencipe. Coordination without communication: The case of the flocking problem. In *Elsevier Science*, 2003.

[Gra59]   P.-P. Grassè. La reconstruction du nid et les coordinations inter-individuelles chez *Bellicositerms natalensis* et *Cubitermes* sp. la thèorie de la stigmergie: essai d'interprètation du comportement des termites constructeurs. In *Insectes Sociaux*, 1959.

[GvBP00]   Enrico H. Gerding, David D. B. van Bragt, and J. A. La Poutre. Scientific approaches and techniques for negotiation. a game theoretic and artificial intelligence perspective. Technical Report SEN-R0005, CWI, 2000.

[GW92]   R.C. Gonzalez and P.A. Wintz. *Digital Image Processing*. Addison Wesley, 1992.

[GWP05]   Andreas Goebels, Alexander Weimer, and Steffen Priesterjahn. Using cellular automata with evolutionary learned rules to solve the online partitioning problem. In *Proceedings of the IEEE Congress on Evolutionary Computation (CEC'05)*, pages 837–843, Edinburgh, 2005. IEEE Press.

[GZ04]   Christian Guttmann and Ingrid Zukerman. Towards models of incomplete and uncertain knowledge of collaborators' internal resources. In *Second German Conference on MultiAgent system TEchnologieS (MATES)*, pages 58–72, Erfurt, Germany, 2004. Springer: LNAI.

[Hah03]   Christian S. Hahn. A detailed analysis of organizational forms for holonic multiagent systems. Diploma thesis, University of Saarland, Saarbrücken, 2003.

[HKD03]   J. Handl, J. Knowles, and M. Dorigo. Ant-based clustering: a comparative study of its relative performance with respect to *k*-means, average link and 1d-som. Technical Report TR/IRIDIA/2003-24, Universite Libre de Bruxelles, 2003.

[HL95]   Bruce Hendrickson and Robert Leland. A multilevel algorithm for partitioning graphs. In *Proc. Supercomputing '95*, San Diego, 1995. ACM Press, New York.

[HM99]   Owen Holland and Chris Melhuish. Stigmergy, self-organization and sorting in collective robots. In *Artificial Life Volume 5, Issue 2*, pages 173–202, Cambridge, MA, USA, 1999. MIT Press.

[Hol75]   J. Holland. *Adaption in Natural and Artificial Systems*. University of Michigan Press, 1975.

[HR95]   M. Heath and P. Raghavan. A cartesian parallel nested dissection algorithm. *SIAM Journal of Matrix Analysis and Applications*, 16(1):235–253, 1995.

[Jon92]   C. A. Jones. *Vertex and Edge Partitions of Graphs*. Phd thesis, Penn State, Dept. Computer Science, State College, PA, 1992.

[JP98]   H. Juille and J. Pollack. Coevolving the ideal trainer: Application to the discovery of cellular automata rules. In *Proceedings of the Third Annual Genetic Programming Conference (GP-98)*, 1998.

[Kaz00]   Sanza Kazadi. Swarm engineering. Phd thesis, California Institute of Technology, 2000.

[KE01]   James Kennedy and Russel C. Eberhart. *Swarm Intelligence*. Morgan Kaufmann, 2001.

[Kem06]   Thomas Kemmerich. Partikelintelligenz und insektenbasierte Algorithmen zur dynamischen Task Allocation. Bachelor thesis, Universität Paderborn, 2006.

[Ken06]   Thobias Kenter. Emergente Organisationsbildung in Multiagentensystemen als Heuristik für ein Optimierungsproblem. Bachelor thesis, Universität Paderborn, 2006.

[Kir05]   Melanie Kirchner. Genetisch gesteuertes Lernen von Regeln für eindimensionale zelluläre Automaten. Bachelor thesis, Universität Paderborn, 2005.

[KJ97]   S. Kalenka and N.R. Jennings. Socially responsible decision making by autonomous agents. In *Cognition, Agency and Rationality*, pages 135–149. Kluwer Academic Publishers, 1997.

[KMN06]   Iva Kozakova, Ronald Meester, and Seema Nanda. The size of components in continuum nearest-neighbour graphs. *Annals of Probability*, 34(2):528–538, 2006.

[KS94]   P. Kuntz and D. Snyers. Emergent colonization and graph partitioning. In *Proceedings of the Third International Conference on Simulation of Adaptive Behaviour: From Animals to Animats 3*, pages 494–500. MIT Press, Cambridge, MA, 1994.

[KS01]   T.J. Koo and S.M. Shahruz. Formation of a group of unmanned aerial vehicles (uavs). In *Proceedings of the American Control Conference*, 2001.

[KSJ02]   Peter Kostelnik, Marek Samulka, and Milan Janosik. Scalable multi-robot formations using local sensing and communication. In *Proceedings of the third International Workshop on Robot Motion and Control (RoMoCo'02)*, Poznan, Poland, 2002.

[KW97]   H. Krapp and T. Wägenbauer. *Komplexität und Selbstorganisation*. SP, München, 1997.

[LB95]   M.W.S. Land and R.K. Belew. No two-state ca for density classification exists. In *Physical Review Letters, 74(25):5148-51*, 1995.

[LG99]   A. E. Langham and P. W. Grant. Using competing ant colonies to solve k-way partitioning problems with foraging and raiding strategies. In *ECAL '99: Proceedings of the 5th European Conference on Advances in Artificial Life*, pages 621–625, London, UK, 1999. Springer-Verlag.

[Lyn96]   Nancy A. Lynch. *Distributed Algorithms*. Morgan Kaufmann, 1996.

[Mae94]   P. Maes. Agents that reduce work and information overload. In *Communication of the ACM 37 No. 7*, pages 31–40, 1994.

[Mat98]   Maja J. Matarić. Using communication to reduce locality in distributed multi-agent learning. In *Journal of Experimental and Theoretical Artificial Intelligence, special issue on Learning in DAI Systems*, pages 357–369. Gerhard Weiss, ed., 10(3), 1998.

[MC96]   Pattie Maes and Anthony Chavez. Kasbah: An agent marketplace for buying and selling goods. In *Proceedings of the First International Conference on the Practical Application of Intelligent Agents and Multi-Agent Technology*, London, UK, 1996.

[MD00]   S. Moss and P. Davidsson. Multi-agent-based simulation (introduction). In *Second international Workshop on Multi-agent-based Simulation (MABS)*, Germany, 2000. Springer.

[MHC93]   Melanie Mitchell, Peter T. Hraber, and James P. Crutchfield. Revisiting the edge of chaos: Evolving cellular automata to perform computations. In *Complex Systems 7*, pages 89–130, 1993.

[Mor96]   R. Morley. Painting trucks at general motors: The effectiveness of a complexity-based approach. In *Embracing Complexity: Exploring the Application of Complex Adaptive Systems to Business*, pages 53–58, Camebridge, MA, USA, 1996.

[MTTV93]   G. Miller, S.-H. Teng, W. Thurston, and S. Vavasis. Automatic mesh partitioning. In *Graph Theory and Sparse Matrix Computation*. Springer-Verlag, 1993.

[Mue96]   J.P. Mueller. The design of intelligent agents: a layered approach. In *Lecture Notes in Artificial Intelligence LNAI 1177, 2nd Print*, Berlin, Germany, 1996. Springer.

[Nii86]   H. Penny Nii. Blackboard systems, part one: The blackboard model of problem solving and the evolution of blackboard architectures. *AI Magazine*, 7(2):38–53, 1986.

[NORL86]   B. Nour-Omid, A. Raefsky, and G. Lyzenga. Solving finite element equations on concurrent computers. *American Soc. Mech. Eng. (Editor A. K. Noor)*, pages 291–307, 1986.

[ORF96]   C. Ou, S. Ranka, and G. Fox. Fast and parallel mapping algorithms for irregular and adaptive problems. *Journal of Supercomputing*, 10:119–140, 1996.

[Par82]   B. L. Partridge. The structure and function of fish schools. In *Scientific American*, pages 114–123, 1982.

[PGW05]   Steffen Priesterjahn, Andreas Goebels, and Alexander Weimer. Stigmergetic communication for cooperative agent routing in virtual environments. In *Proceedings of the International Conference on Artificial Intelligence and the Simulation of Behaviour (AISB2005)*, Hatfield, Great Britain, 2005.

[PKea03]   D. W. Palmer, M. Kirschenbaum, and et al. Using a collection of humans as an execution testbed for swarm algorithms. In *IEEE Swarm Intelligence Symposium*, pages 58–64, 2003.

[PKWG05]   S. Priesterjahn, O. Kramer, A. Weimer, and A. Goebels. Evolution of reactive rules in multi player computer games based on imitation. In *Proceedings of the International Conference on Natural Computing (ICNC'05)*, pages 744–755, Changsha, China, 2005. Springer, LNCS 3611.

[PKWG06]   S. Priesterjahn, O. Kramer, A. Weimer, and A. Goebels. Evolution of human-competitive agents in modern computer games. In *Proceedings of the IEEE World Congress on Computational Intelligence (WCCI'06)*, Vancouver, Canada, 2006.

[PL03]   Liviu Panait and Sean Luke. Cooperative multi-agent learning: The state of the art. In *Technical Report GMU-CS-TR-2003-1*. George Mason University, USA, 2003.

[PM01]   L. Pagie and M. Mitchell. A comparison of evolutionary and coevolutionary search. In *R.K. Belew and H. Juillé (editors): Coevolution: Turning Adaptive Algorithms upon Themselves*, pages 20–25, USA, 2001.

[Rei65]   F. Reif. *Fundamentals of Statistical and Thermal Physics*. McGraw-Hill, 1965.

[Rey87]   Craig W. Reynolds. Flocks, herds and schools: A distributed behaviour model. In *Proceedings of SIGGRAPH'87. Reprinted in Computer Graphics 21 (4)*, pages 25–34, 1987.

[RPH94]   G. E. Robinson, R. E. Page, and Z.-Y. Huang. Temporal polyethism in social insects is a developmental process. *Animal Behaviour*, 48:467–469, 1994.

[RZ94]   J.S. Rosenschein and G. Zlotkin. Designing conventions for automated negotiation. In *AI Magazine*, pages 29–46, 1994.

[Sch03]   M. Schillo. Self-organization and adjustable autonomy: Two sides of the same medal? *Connection Science*, 4(14):345–359, 2003.

[Sch04]   M. Schillo. *Multiagent Robustness: Autonomy vs. Organisation*. Phd thesis, Department of Computer Science, Universität des Saarlandes, Germany, 2004.

[SCW02]   Wei-Min Shen, Cheng-Ming Chuong, and Peter Will. Simulating self-organization for multi-robot systems. In *International Conference on Intelligent and Robotic Systems*, Switzerland, 2002.

[See95]   T.D. Seeley. *The Wisdom of the Hive*. Harvard University Press, London, 1995.

[SF03]   M. Schillo and K. Fischer. Holonic multiagent systems. *KI Zeitschrift*, 4(54):327–332, 2003.

[SFF⁺04]   M. Schillo, K. Fischer, B. Fley, M. Florian, F. Hillebrandt, and D. Spresny. FORM - a sociologically founded framework for designing self-organization of multiagent systems. In *Proceedings of the International Workshop on Regulated Agent-Based Social Systems: Theories and Applications*, pages 156–175. Springer-Verlag, LNAI 2934, 2004.

[SFS03]   Michael Schillo, Klaus Fischer, and Jörg Siekmann. The link between autonomy and organisation in multiagent systems. In *Proceedings of the First International Conference on Applications of Holonic and Multiagent Systems (HoloMAS'03)*, pages 81–90, Berlin et. al., Germany, 2003. Lecture Notes in Artificial Intelligence, vol. 2744, Springer.

[She03]   Dylan A. Shell. An annotated bibliography of papers that make use of the word stigmergy. robotics.usc.edu/d̄shell/res/annot.pdf, 2003.

[Sik01]   Axel Sikora. *Wireless LAN - Protokolle und Anwendungen*. Addison-Wesley, 2001.

[SKK00]   Kirk Schloegel, Georg Karypis, and Vipin Kumar. Graph partitioning for high performance scientific simulations. Technical Report TR 00-018, Department of Computer Science and Engineering, University of Minnesota, Minneapolis, 2000.

[Smi76]   Adam Smith. *An Inquiry into the Nature and Causes of the Wealth of Nations*. Methuen and Co., Ltd., 1776.

[SS96]   K. Sugihara and I. Suzuki. Distributed algorithms for formation of geometric patterns with many mobile robots. *Journal of Robotic Systems 13*, 3:127–139, 1996.

[SS98]   Frank Schweitzer and Gerald Silverberg, editors. *Evolution and Self-Organization in Economics*, Selbstorganisation - Jahrbuch für Komplexität in den Natur-, Sozial- und Geisteswissenschaften (9). Duncker & Humbold, Berlin, 1998.

[SS04]   Erol Sahinn and William M. Spears, editors. *Swarm Robotics*, Lecture Notes in Computer Science, LNCS 3342. Springer ISSN 0302-9743, 2004.

[ST97]   Horst D. Simon and Shang-Hua Teng. How good is recursive bisection? *SIAM J. Sci. Comput.*, 18(5):1436–1445, 1997.

[SV00]   Peter Stone and Manuela M. Veloso. Multiagent systems: A survey from a machine learning perspective. In *Autonomous Robots 8 (3)*, pages 345–383. Springer, 2000.

[SVZ04]   Christian Schindelhauer, Klaus Volbert, and Martin Ziegler. Spanners, weak spanners and power spanners for wireless networks. In *Proceedings of ISAAC 2004*, pages 805–821. LNCS 3341, Springer, 2004.

[SW03]   S. Schambeger and J.M. Wierum. Graph partitioning in scientific simulations: Multilevel schemes versus space-filling curves. In *Proceedings of the 7th International Conference on Parallel Computing Technologies*, volume 2763 of *LNCS*, pages 165–179, 2003.

[SWC04]   A. J. Soper, C. Walshaw, and M. Cross. A combined evolutionary search and multilevel optimisation approach to graph-partitioning. In *Journal of Global Optimization 29*, pages 225–241, 2004.

[SY99]   Ichiro Suzuki and Masafumi Yamashita. Distributed anonymous mobile robots: Formation of geometric patterns. *SIAM Journal on Computing*, 28(4):1347–1363, 1999.

[TA04]   Predrag T. Tosic and Gul A. Agha. Maximal clique based distributed coalition formation for task allocation in large-scale multi-agent systems. In *Proceedings of the first Int. Workshop on Massively Multi-Agent Systems (MMAS 2004)*. Springer, LNAI 3446, 2004.

[TBD98]   G. Theraulaz, E. Bonabeau, and J. Deneubourg. Response threshold reinforcement and division of labour in insect societies. *Proc. Roy. Soc.*, pages 327–332, 1998.

[Tin05]   Chuan-Kang Ting. On the convergence of multi-parent genetic algorithms. In *Proceedings of the 2005 Congress on Evolutionary Computation*, pages 396–403, Edinburgh, UK, 2005. IEEE Press.

[Ü93]   C. Ünsal. Self-organization in large populations of mobile robots. Master thesis, Virginia Polytechnic Institute and State University. Blacksburg, Virginia, 1993.

[Vai89]   P. Vaidya. An $O(n \log n)$ algorithm for the all-nearest-neighbors problem. In *Discrete and Computational Geometry (4)*, pages 101–115, 1989.

[Wei99]   G. Weiss. *Multiagent Systems - A modern approach to Distributed Artificial Intelligence*. The MIT Press, Cambridge, Massachusetts, 1999.

[WF05]   Horst F. Wedde and Muddassar Farooq. Beehive: Routing algorithms inspired by honey bee behaviour. *KI Künstliche Intelligenz*, 4/05(4):18–24, 2005.

[WHN05]   A. Winfied, C.J. Harper, and J. Nembrini. Towards dependable swarms and a new discipline of swarm engineering. In *Swarm Robotics: State-of-the-Art Survey*, pages 126–142. Springer, LNCS, 2005.

[Wik06]   Wikipedia. Swarm robotics — wikipedia, the free encyclopedia, 2006. [Online; accessed 15-June-2006].

[Wil84]   E. O. Wilson. The relation between caste ratios and division of labour in the ant genus phedoile (hymenoptera: Formicidae). *Behavioural Ecology and Sociobiology*, 16:89–98, 1984.

[WL02]   Y. Wang and X.-Y. Li. Distributed spanner with bounded degree for wireless ad hoc networks. In *Parallel and Distributed Computing Issues in Wireless Networks and Mobile Computing*, 2002.

[WMC00]   J. Werfel, M. Mitchell, and J.P. Crutchfield. Resource sharing and coevolution in evolving cellular automata. In *IEEE Transactions on Evolutionary Computation, 4(4):388*, 2000.

[Wol02]   Stephen Wolfram. *A New Kind of Science*. Wolfram Media, 2002.

[Woo02]   Michael J. Wooldridge. *An introduction to multiagent systems*. Wiley, Chichester, reprint edition, 2002.

[WPG05]   Alexander Weimer, Steffen Priesterjahn, and Andreas Goebels. Towards the emergent memetic control of a module robot. In *Proceedings of the International Conference on Artificial Intelligence and the Simulation of Behaviour (AISB2005)*, Hatfield, Great Britain, 2005.

[YJG62]   M.C. Yovits, G.T. Jacobi, and G.D. Goldstein. *Self-Organizing Systems*. McGregor & Werner, Washington D.C., 1962.

[Yu00]   Lei Yu. *Agent Oriented and Role Based Business Process Management*. Phd thesis, University of Zürich, 2000.

[ZS06]   Robert Zlot and Anthony Stentz. Market-based multirobot coordination for complex tasks. *Intl. J. of Robotics Research, Special Issue on the 4th Intl. Conference on Field and Service Robotics*, 25(1), 2006.

Index